Flavouring and Fragrant Resources of India

S.K. Sood
Parul Gupta
Suresh Kumar
Department of Biosciences,
Himachal Pradesh University,
Summer Hill, Shimla - 171 005 (HP)

SCIENTIFIC
PUBLISHERS (INDIA)
www.scientificpub.com

Published by:
Scientific Publishers (India)
5-A, New Pali Road, P.O. Box 91
JODHPUR - 342 001
e-mail: info@scientificpub.com
www.scientificpub.com

Print : 2020

ISBN: 978-81-7233-635-6

Lasertype set: Rajesh Ojha

Binding at: Rajasthan Law Book Binding Works, Jodhpur

Printed in India

PREFACE

Present volume is an endeavour towards bringing together widely scattered and diffused information on ethnic Indian flavouring and fragrant plants from disparate sources to form a highly structured work of reference to the world at large. As many as 625 species of plants belonging to 331 genera under 121 families have been enlisted alphabetically by genera and species, of which 3 species (3 genera) belong to Algae, 23 species (11 genera) to Lichens, 3 species (2 genera) to Pteridophytes, 16 species (9 genera) to Gymnosperms and 580 species (356 genera) to Angiosperms. For easy comprehension by laymen and specialists alike, the data given in each species include correct botanical name, synonyms, family, English, Sanskrit and Hindi names, distribution, part/s used, procedural details of utilisation, active principles and the original source of information for authenticity in most cases. By and large, the flavouring and fragrant plants belong to Lamiaceae (56 species), Apiaceae (35 species), Zingiberaceae (35 species), Asteraceae (31 species), Rutaceae (28 species), Rosaceae (27 species), Liliaceae (19 species), Lauraceae (17 species), Myrtaceae (14 species), Poaceae (13 species), Brassicaceae (12 species), etc. For familiarizing the reader with the flavouring and fragrant plants, over 46 colour photographs have been included in the text. It is earnestly hoped that the synthesis provided will certainly bridge the gap between the traditional users and scientific fraternity and also it can be put to use for commercial exploitation in the country.

This venture could not have been possible without the valuable advice and incessant encouragement of Prof. S.K. Gupta, Vice-Chancellor, Himachal Pradesh University, Shimla, for which we express our sincere gratitude to him. Well deserved appreciation is due to Prof. D.C. Kalia and Dr. V.K. Mattu and other colleagues of Department of Biosciences for their co-operation and well wishes; to Mr. Dhiraj Rawat, Ms. Anjna and Mr. Munish for their ungrudging help in critically reading the various drafts of this manuscript, and to Mr. Bal Krishan for his help in various ways. Thanks are also due to authors of books, papers and figures listed in the compendium. Finally, we would like to thank the members of our families for moral support and unflinching patience during the course of the work.

Suggestions for further improvement of the compendium will always be welcomed and entertained whole heartedly.

S.K. Sood
Parul Gupta
Suresh Kumar

ABBREVIATIONS

α	Alpha
alt.	Altitude
•	Beta
C	Central
•	Delta
de-O	Deoxy
D	Dextro
E.	East/Eastern
Eng.	English
•	Epsilon
et al.	et alia (and other authors)
etc.	et cetera (so on)
Et	Ethyl
•	Gamma
H.P.	Himachal Pradesh
H	Hindi
OH	Hydroxy
i.e.	id est (that is)
L	Laevo
M	Meta
m	Meter
Me	Methyl
mg	Milligram
N	North/Northern
N.E.	North-East
N.W.	North-West
P	Para
%	Per cent
Pl. (s)	Plate (s)
Sans.	Sanskrit
S	South/Southern
S.E.	South-East
S.W.	South-West
sp	Specie
spp	Species
Syn.	Synonym
var.	Variety
Vit.	Vitamin
W.	West/Western

CONTENTS

Plate 1

Plate 1 (A-F) Flavouring and Fragrant Plants of Ethnic India
A. *Abelmoschus esculentus* (Linn.) Moench; **B.** *Aegle marmelos* Correa ex Roxb.; **C.** Plant of *Allium cepa* Linn.; **D.** Bulbs of *Allium cepa* Linn.; **E.** Plant of *Allium sativum* Linn.; **F.** Bulbs of *Allium sativum* Linn.

Plate 2

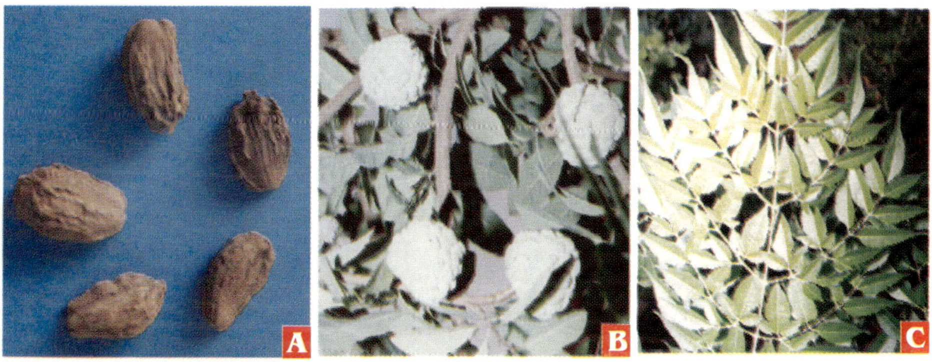

Plate 2 (A-C) Flavouring and Fragrant Plants of Ethnic India
A. *Amomum subulatum* Roxb.; **B.** *Annona squamosa* Linn.;
C. *Azadirachta indica* A. Juss.

Plate 9

Plate 9 (A-C) Flavouring and Fragrant Plants of Ethnic India.
A. *Nigella sativa* Linn.; **B.** *Nyctanthes arbor-tristis* Linn.; **C.** *Ocimum sanctum* Linn.

Plate 10

Plate 10 (A-E) Flavouring and Fragrant Plants of Ethnic India

A. *Pandanus odoratissimus* Linn. f.;
B. *Prunus amygdalus* Batsch;
C. *Punica granatum* Linn.;
D. *Rosa* sp.;
E. *Santalum album* Linn.

Plate 11

Plate 11 (A-D) Flavouring and Fragrant Plants of Ethnic India.
A. *Sesamum indicum* Linn.; **B.** *Shorea robusta* Gaertn. f.; **C.** *Syzygium aromaticum* (Linn.) Merrill and Perry; **D.** *Tamarindus indica* Linn.

1

WHAT ARE ESSENTIAL OILS?

Essential oils are a mixture of odorous compounds which are products of secondary metabolism of plants and isolated with minimum chemical changes from human interventions. These evaporate in contact with the air and possess a pleasant fragrance and, therefore, also known as volatile oils. Essential oils are isolated by steam distillation from many different types of plant materials: wood (sandal wood), bark (cinnamon), leaves (patchouli), stem (ginger), root (vetiver), flower bud (clove), stigmas (saffron), flowers (jasmine, rose), fruits (orange), and seeds (nutmeg), at concentrations ranging from thousandths of a per cent to one or several per cent. Generally, the oils are contained in specialised secretory structures which include secretory cells, ducts, cavities, glandular trichomes, etc. Most aromatic plants (highly fragrant plants) produce essential oils from glandular cells throughout the plant. Chemically, the predominant constituents of volatile oils are mono- and sesqui-terpene hydrocarbons and oxygenated materials biogenically derived from them, phenyl propanoids and alkane derivatives (Hegnauer, 1982). As many as 3000 essential oils have been identified from the 87 families of the vegetal kingdom. Most of the essential oil yielding plants belong to Lamiaceae, Apiaceae, Lauraceae, Myrtaceae, Rosaceae, Rutaceae, Asteraceae, Geraniaceae, Poaceae, Fabaceae, etc. Over 200 components may be present in an essential oil and the characteristic aromatic profile may depend on trace amounts of one or several compounds (Ram and Ram, 1989). Traditionally, they are used for flavouring and aroma in foods and beverages (baking products; frozen foods; meat, poultry and fish products; desserts and puddings; soups, sauces and pickles; soft and hard drinks; candy and chocolates, etc.), but they also add fragrance to cosmetic products, disguise the taste of medicines, and flavour liquors and tobacco products.

2

SPICES AND CONDIMENTS

In general, all aromatic or pungent vegetable products that are used more for their aromatic flavour than for their food value are known as spices. This word is derived from the latin word species, meaning specific kind, and later, goods of merchandise. The quest for condiments and spices was once a powerful force in world history. The search for spices and the wealth and power that came with them encouraged the great explorers, Columbus, Magellan and Da Gama to venture on journeys of exploration in search of a new route to the valuable commodities of the East in the middle ages. According to the American Spice Trade Association (ASTA), spices are 'any dried plant product used primarily for seasoning purposes'. This definition covers a wide range of plant parts: tropical aromatics, leafy herbs, spice seeds, roots, dehydrated vegetables, and spice blends. Condiments, on the other hand, are spices or other flavouring substances which possess a sharp taste, and are commonly added to food after it has been cooked. However, according to the International Organization Standardization (ISO), there is no clear-cut division between 'spices' and 'condiments' and as such they have been clubbed together (Pruthi, 1976). Typical examples of spices are cloves (buds), cinnamomum (bark), turmeric (rhizome), peppercorns (berries), vanilla (the bean from a tropical orchid vine) and cumin, coriander, dill and fennel (seeds), etc. Spices can be grouped into five basic categories. These are: Sweet: cinnamon, allspice, nutmeg, vanilla; Pungent: cloves, star anise, cardamom; Tangy: ginger, tamarind, sumach, kokam; Hot: pepper, chilli, mustard, horse radish; amalgamating: coriander seed, fennel seed. Most of the herbs, such as thyme, sage, marjoram, oregane, bay leaves, mint and rosemary are referred to as savoury. The herbs do have varying degrees of flavour intensity, however not as dramatic as with spices. Of the total world trade in spices, India alone contributes to 25-30 per cent. The major spices grown in India are pepper, cardamom, ginger, turmeric and chillies, and their export help in earning valuable foreign exchange.

3

OVERVIEW OF THE WORK ON FLAVOURING AND FRAGRANT PLANTS

Notable works on flavouring and fragrant plants of the world are those of Winton and Winton (1939) -*'Structure and Composition of Foods'*; Verill (1940) -*'Perfumes and Spices'*; Guenthor (1948-52)-*'The Essential Oils'*; Arctander (1960) -*'Perfume and Flavour Materials of Natural Origin'*; Rosemary (1966) - *'The Penguin Book of Herbs and Spices'*; Loewenfeld and Back (1974)- *'The Complete Book of Herbs and Spices'*; Poucher (1974) - *'Perfumes, Cosmetics and Soaps'*; Devon and Scott (1975) -*'Handbook of Naturally Occurring Compounds'*; Genders (1977)-*'Scented Flora of the World'*; Charalambons and Inglett (1978)-*'Flavour of Foods and Beverages: Chemistry and Technology'*; Heath (1978, 1981) -*'Flavour Technology: Profiles, Products, Applications'*; *'Source Book of Flavours'*, respectively; and Lawless (1995)- *'The Illustrated Encyclopedia of Essential Oils'*. Nevertheless a few valuable contributions in this regard from India have been made by Badwhar *et al.* (1964) -*'Some Useful Aromatic Plants'*; Pruthi (1976, 1987, 1991a-d, 1992) - *'Spices and Condiments'*; *'Bibliography on 57 Spices and Condiments (1974-86)'*; *'Major Spices of India'*; *'Minor Spices and Condiments'*; *'Modern Methods of Analysis of Spices and Spice Products'*; *'Advances (Innovations) in Spice Processing Technology'*; *'Post-Harvest Technology of Spices and Condiments'*, respectively; Atal and Kapur (1982) - *'Cultivation and Utilization of Aromatic Plants'*; Chadha (1982)-*'Flavours and Their Industrial Applications'*; Hussain *et al.* (1988)- *'Major Essential Oil-Bearing Plants of India'*; Ram and Ram (1989)-*'Fragrant Plants'*; Mehtani and Ingle (1999) -*'Plant Food Flavours'*; Shiva *et al.* (2002)- *'Aromatic and Medicinal Plants'*; and Baskar Rajan *et al.* (2003) - *'Spices Medicine'*. In spite of a plethora of literature available on the relevant theme, no comp-rehensive work has been done so far on them in India (Pruthi, 1987). The present work is a solitary attempt in this direction and aims at synthesising the current state of our folk-lore knowledge on the subject.

4

PRESENTATION AND ARRANGEMENT OF DATA

Practically, the methodology followed in this work is the same as that described by Jain (1964, 1967), involving compilation of information from the earlier published scientific literature and recent articles on flavouring and fragrant plant species used by the natives from different parts of country in different journals on economic botany and ethnobotany. The plant species are arranged in alphabetical order each by its correct Botanical name (nomenclature based on Bennett, 1987) and its synonyms along with their respective families, followed by English, Sanskrit and Hindi Names; Distribution; Part's Used; Utilization; Active Principles and relevant References, if known. Altogether, 48 colour photographs have been included to explain the subject-matter. Besides, index to total number of genera and species of flavouring and fragrant plants under various divisions of plant kingdom (Appendix I), families of flavouring and fragrant plants (Appendix II), flavouring and fragrant species under a genus (Appendix III), index to botanical names (Appendix IV), index to English names (Appendix V), index to Sanskrit names (Appendix VI) and index to Hindi names (Appendix VII) have also been included as appendices.

5

FLAVOURING AND FRAGRANT PLANTS

1. ***Abelmoschus esculentus*** (Linn.) Moench (Pl. 1A)

Syn : *Hibiscus esculentus* Linn.; *H. longifolius* Roxb.

Family : Malvaceae.

English, Sanskrit and Hindi Names : **Eng. :** Edible Hibiscus, Gobba, Gumbo, Lady's finger, Ochro, Okra; **Sans.:** Asrapatraka, Bhenda, Chatupunda, Darivka, Sushaka; **Hindi:** Bhindi, Bhindi Tori, Katavandai, Ramturai, Ranturi.

Distribution : Cultivated throughout India and all other tropical countries.

Part/s Used : Seeds.

Utilization : Ripe seeds roasted for use as substitute for coffee. A vegetable gum, called Okra gum, obtained from the plant used for flavouring and bodying agent in vegetable soups and gravies.

Active Principles : Okra-mucilage from seeds rich in β-sitosterol and its 3β-galactoside.

References : Ambasta (1986), Asolkar *et al.* (1992), Bennett (1987), Kirtikar and Basu (1984), Rastogi and Mehrotra (1990, 1995), Watt (1972).

2. ***Abelmoschus ficulneus*** Wight and Arn.

Syn : Hibiscus ficulneus Linn.; *H. strictus* Roxb.; *H. prostratus* Roxb.

Family : Malvaceae.

English and Hindi Names : **Eng. :** White Wild Musk Mallow; **Hindi :** Ran bhendi.

Distribution : Indigenous in the hotter parts of India from Punjab to Madras.

Part/s Used : Seeds.

Utilization : Aromatic seeds used for perfuming coffee and sweetmeats.

Active Principles : Seed oil yields epoxyoleic (4.9), malvalic (4.2) and sterculic (1.0%) acids.

References : Ambasta (1986), Bennett (1987), Chandra (1997), Mehtani and Ingle (1999), Rastogi and Mehrotra (1993), Watt (1972).

3. ***Abelmoschus moschatus*** Medic.

Syn : *Hibiscus abelmoschus* Linn.; *H. chinensis* Wall.; *H. flavescens* Cav.; *H. ricinifolius* Wall.; *A. rugosus* Wight and Arn.; *H. sagittifolius* Kurz.; *H. spathaceus* Wall.

Family : Malvaceae.

English, Sanskrit and Hindi Names: **Eng.:** Ambrette plant, Musk mallow; **Sans.:** Latakasturikam; **Hindi:** Kasturidana, Muskdana.

Distribution : Deccan. Foothills of Himalaya. Hotter parts of India. Cultivated in the lower hills of H.P.

Part/s Used : Seeds. Flowers.

Utilization : Flowers much in demand for flavouring tobacco and seeds for flavouring coffee. Seeds yield 0.2 – 0.6% volatile oil which smells like musk and used mainly in fine fragrances. The oil is also used to flavour non-alcoholic beverages, confectionery, baked goods, gelatins and puddings.

Active Principles : These are trans-2-trans-b-farnesyl acetate (64.22%), and ambrettolide (14.96%), (2E, 6E) – farnesyl acetate, 1, 16 – (Z) – hexa decene -- 7- olide, (2Z, 6E) – farnesyl acetate, (2E, 6E) – farnesol, decyl acetate, dodecyl acetate, (2E, 6E) – farnesyl propionate, 1,14 – (Z) – tetradecene – 5 – olide, and (2E, 6E)-farnesal (essential oil), β-sitosterol, its β-D-glucoside, myricetin and its glucoside (petals).

References : Ambasta (1986), Bennett (1987), Chandra (1997), Harborne and Baxter (2001), Kirtikar and Basu (1984), Mehtani and Ingle (1999), Mittre (1981), Rastogi and Mehrotra (1990, 1991, 1993, 1995), Sharma and Sood (1997), Watt (1972).

4. ***Acacia concinna*** DC.

Syn : *A. rugata* (Lamk.) Merr.; *Mimosa concinna* Willd. Roxb.

Family : Mimosaceae.

Hindi Names: **Hindi :** Kochi, Ritha.

Distribution : Common in the tropical jungles throughout India.

Part/s Used : Leaves.

Utilization : Tender acidic leaves used in chutneys and considered as a substitute of tamarind.

Active Principles : Tartaric, oxalic, succinic and ascorbic acids along with alkaloids-calycotomine and nicotine from leaves.

References : Ambasta (1986), Rastogi and Mehrotra (1991), Watt (1972).

5. ***Acacia dealbata*** Link.

Family : Mimosaceae.

English Name: **Eng.** : Silver Wattle.

Distribution : Nilgiris.

Part/s Used : Flowers.

Utilization : Flowers used for the preparation of Mimosa perfume.

Active Principles : A chalcone – 2', 4, 4', 6'- tetrahydroxy – 3 –methoxy-chalcone -2',-O- β -D-glucoside (I)-along with cernuoside isolated from flowers.

References : Ambasta (1986), Watt (1972).

6. ***Acacia farnesiana*** Willd.

Family : Mimosaceae.

English, Sanskrit and Hindi Names : **Eng. :** Cassie flowers, Sponge tree, Stinking Acacia; **Sans.:** Arimeda, Girimeda, Irimeda, Rimeda, Vita; **Hindi :** Gandbabul, Pissibabul, Vilayati Kikar.

Distribution : Naturalised throughout India, often planted.

Part/s Used : Flowers. Leaves.

Utilization : Flowers source of sweet perfume, called Cassie perfume, and used occasionally in flavours. Tender leaves used for flavouring chutneys.

Active Principles: Isorhamnetin-3,7-glucorhamnoside isolated from flowers. The characteristic fragrance of the Cassie oil has been attributed to cis-3-methyldec-3-en-1-ol, cis-3-methyldec-ene-3-oic acid and trans-3-methyl-dec-4-enoic acid.

References : Ambasta (1986), Kirtikar and Basu (1984), Mehtani and Ingle, Raju (2000), Watt (1972).

7. ***Acacia jacquemontii*** Benth.

Family : Mimosaceae.

Distribution : Ascending to 850 m on the outer Himalaya. Common in ravines and dry water courses in Rajputana and North Gujarat.

Part/s Used : Flowers. Root-bark.

Utilization : Flowers sweetly scented. Root-bark added to the fermented solution of Mahuwa flowers (broth) to flavour the local spirit distilled from it.

References : Asolkar *et al.* (1992), Chandra (1997), Chopra *et al.* (1956), Singh and Pandey, Watt (1972).

8. *Achillea millefolium* Linn.

Syn : A. lanulosa Nutt.

Family : Asteraceae.

English and Hindi Names : **Eng. :** Arrowroot, Bloodwort, Camil, Carpenter Grass, Devil's Nettle, Dog Daisy, Milfoil, Thousand leaved clover, Tansy, Yarrow; **Hindi :** Gandana.

Distribution : W. Himalaya-Kashmir to Kumaon: 2,000 – 3,000m.

Part/s Used : Aerial Parts.

Utilization : Aromatic aerial parts used for flavouring salads, soups and fish. Aromatic property strongest in the flowers.

Active Principles : Three main constituents of volatile oil extracted from flowers are Ivaol, Moschatin and Achilleine. 8–10% cineole is also present.

References : Ambasta (1986), Asolkar *et al.* (1992), Bhattacharjee (1997), Chandra (1997), Chevallier (1996), Chopra *et al.* (1980), Kaul (1997), Kirtikar and Basu (1984), Rastogi and Mehrotra (1990, 1991), Sharma and Sood (1997), Watt (1972).

9. *Achras zapota* Linn.

Syn : Achras sapota Linn.; *Manilkara zapota* (L.) P.van Royen; *Mimusops manilkara* Don.

Family : Sapotaceae.

English and Hindi Names : **Eng. :** American bully, Australian Cranberry, Bully tree, Mazeberry, Neesberry, Rough chaff, Sapodilla plum, Sapota plum; **Hindi :** Sapota.

Distribution : Cultivated more or less throughout India.

Part/s Used : Fruits.

Utilization : Fruits used for flavouring purposes.

Active Principles : Glucosides, tannins, saponins and alkaloids present.

References : Ambasta (1986), Chopra *et al.* (1956), Kirtikar and Basu (1984), Rastogi and Mehrotra (1991), Watt (1972).

10. *Acorus calamus* Linn.

Family : Araceae.

English, Sanskrit and Hindi Names: **Eng. :** Sweet flag; **Sans. :** Bhuta – nashini, Galani, Jatila, Ugragandha, Vacha; **Hindi :** Bach, Ghorabach, Safed – bach.

Distribution : Damp marshy places in various parts of India from 900 – 1,800m. Cultivated in Karnataka and other regions.

Part/s Used : Leaves. Rhizomes.

Utilization : The Calamus oil obtained from rhizome has spicy fragrance and used in perfumery and also to flavour alcoholic beverages. Lemony smelling leaves yield essential oil.

Active Principles : The aromatic volatile oil isolated from dry rhizome contains asaryl aldehyde, an amorphous bitter principle, acorin (glucoside), asamyl alcohol, eugenol, asarone, starch, tannins, mucilage and saponins. Fresh leaves contain oxalic acid 0.078% and calcium 0.006%.

References : Anonymous (1999), Asolkar *et al.* (1992), Bhattacharjee (1998), Chevallier (1996), Greenish (1999), Harborne and Baxter (2001), Jain (1968), Kaul (1997), Kirtikar and Basu (1984), Kumar and Dwivedi (2000), Purohit (2000), Rastogi and Mehrotra (1990, 1991, 1993, 1995), Sharma and Sood (1997), Watt (1972).

11. *Acronychia pedunculata* (L.) Miq.

Syn : Acronychia barberi Gamble; *A.laurifolia* Blume; *Cyminosa pedunculata* DC; *Jambolifera pedunculata* L.

Family : Rutaceae.

English Names : **Eng. :** Claw flowered laurel, Laka wood.

Distribution : Met with in Konkan, North Kanara, W. Ghats, Southern Deccan slopes, Orissa, Sikkim, Khasia Hills, Assam.

Part/s Used : Leaves.

Utilization : Tender leaves have the flavour of cumin, used in salads and also as a condiment to flavour curries.

Active Principles : Kokusaginine isolated from leaves.

References : Ambasta (1986), Bennett (1987), Chandra (1997), Kirtikar and Basu (1984), Mittre (1981), Rastogi and Mehrotra (1993).

12. *Adansonia digitata* Linn.

Syn : Baobabus digitata O. Ktze Rev.

Family : Bombacaceae.

English, Sanskrit and Hindi Names : **Eng. :** African Calabash, Baobab, Cream of Tartar tree, Monkey Broad Tree, Sour Gourd; **Sans. :** Chitrala, Dirghadandi, Gorakshi, Sarpadandi; **Hindi :** Gorakh amli, Gorakh imli, Goramlichora, Hathikathian.

Distribution : Cultivated in some parts of India; common in Andhra Pradesh.

Part/s Used : Leaves.

Utilization : The fresh tender leaves used as a mucilaginous condiment and seasoner for all kinds of food.

Active Principles : Leaves yield bauerenol, friedelin, lupeol, scopo-letin, β-sitosterol, acetates of bauerenol, lupeol and taraxerone.

References : Chandra (1997), Kirtikar and Basu (1984), Rastogi and Mehrotra (1995), Watt (1972).

13. ***Aegle marmelos*** Correa *ex* Roxb. (Pl. 1B)

Syn : Crataeva marmelos Linn.; *C. religiosa* Ainslie.

Family : Rutaceae.

English, Sanskrit and Hindi Names : **Eng. :** Bael fruit tree, Bengal Quince, Golden apple, Holy fruit, Indian Quince, Stone apple; **Sans.:** Adhararuha, Asholam, Bilva, Duraruha, Hridyagandha, Sadaphala, Sriphala; **Hindi :** Bel, Bela, Bili, Sriphal.

Distribution : Wild in the Sub-Himalayan tract, C. and S. India; often planted all over India.

Part/s Used : Flowers. Pulp of Fruit.

Utilization : Fragrant flowers on distillation yield a sweet smelling scent. Ripe fruit's pulp aromatic, sweet and cooling and is commonly used for the preparation of jam, candy, syrup, etc.

Active Principles : Polysaccharides isolated from fruit pulp yielded on hydrolysis galactose (20.4), arabinose (10.7), uronic acid (25.2%), trace of L-rhamnose, xanthotoxol, scoparone, scopo-letin, umbelliferone, marmesin, skimmin β-sitosterol- β-D-glucoside, an essential oil containing d-α-phellandrene, allo-imperatorin and β - sitosterol.

References : Ambasta (1986), Arora (1981), Anonymous (1999), Bajpayee and Dixit (1996), Bhattacharjee (1998), Chevallier (1996), Chopra *et al.* (1980), Greenish (1999), Jain (1968), Kirtikar and Basu (1984), Negi and Gaur (1994), Patole and Jain (2002), Raju (2000), Rastogi and Mehrotra (1990, 1991, 1995), Singh (1999), Sinha and Sinha (2001), Vihari (1995), Watt (1972).

14. ***Aeolanthus myrianthus*** Baker

Syn : Aeolanthus gamwellae Taylor

Family : Lamiaceae.

Distribution : Introduced into India; cultivated in Jorhat, Assam.

Part/s Used : Flowers.

Utilization : Essential oil extracted from the flowers - Oil of Ninde, is used in soaps and perfumery.

Active Principles : Flowers rich in geraniol.

References : Ambasta (1986), Asolkar *et al.* (1992).

15. ***Aframomum melegueta*** (Rosc.) K. Schum.

Syn : Amomum melegueta Rosc.

Family : Zingiberaceae.

English Names: **Eng.:** Alligator Pepper, Grains of Paradise, Meleguetta pepper.

Distribution : Cultivated in Indian gardens.

Part/s Used : Seeds.

Utilization : Seeds give pungency to alcoholic liquors and cordials; and also provide a hot and peppery condiment which is useful as a substitute for pepper.

Active Principles : These are paradol (a pungent principle related to gingerol in ginger) tannins (seeds), humulene (31.3), its epoxide (27.7), β-caryophyllene (8.5), its epoxide (17.9), β-pinene and linolool (seed oil).

References : Ambasta (1986), Asolkar *et al.* (1992), Chevallier (1996), Greenish (1999), Harborne and Baxter (2001), Rastogi and Mehrotra (1995), Watt (1972).

16. *Aganosma caryophyllata* (Roxb. *ex* Sims.) G. Don

Syn : A.dichotoma (Roth.) K. Schum.; *Echites caryophyllata* Wall. non Roxb.

Family : Apocynaceae.

Distribution : Lower Bengal. Risikund. Deccan Peninsula.

Part/s Used : Flowers.

Utilization : Flowers fragrant.

Active Principles : Ferulic, ursolic and vanillic acids, rutin, hyperin, isoquercetin, quercetin-3-arabinoside and β-sitosterol isolated from flowers.

References: Rao & Henry (1995), Rastogi & Mehrotra (1995), Watt (1972)

17. *Ageratum conyzoides* L.

Family : Asteraceae.

English Names : **Eng. :** Goat weed, Floss flower.

Distribution : Found as a weed in different parts of India.

Part/s Used : Flowers.

Utilization : Flowers fragrant.

Active Principles : Flowers contain γ-cadinene, caryophyllene, de-OMe-ageratochromene, ageratochromene and precocene I and II.

References : Asolkar *et al.* (1992), Bhattacharjee (1998), Kirtikar and Basu (1984), Rastogi and Mehrotra (1995).

18. *Aisandra butyracea* (Roxb.) Baehni

Syn : Bassia butyracea Roxb.; *Diploknema butyracea* (Roxb.) Lamk.; *Madhuca butyracea* (Roxb.) Macbr.

Family : Sapotaceae.

English and Hindi Names: **Eng.** : Hill Mahua, Indian Butter Tree; **Hindi** : Phalwara.

Part/s Used : Seeds.

Utilization : Seed oil used for making sweet scented hair oils.

Active Principles : Ethanolic extract of nut kernel contains flavonoid mixture of quercetin, 10%; and dihydroquercetin 90%.

References : Asolkar *et al.* (1992), Negi and Gaur (1994), Raju (2000).

19. *Alliaria petiolata* (Bieb.) Cavara and Grande

Syn : Alliaria officinalis Andrzej *ex* Bieb.; *Arabis petiolata* Bieb.; *Sisymbrium alliaria* Scop.

Family : Brassicaceae.

English Names: **Eng.** : Beggarman's Oatmeal, Cardiacke, Caspere, Eileber, English Treacle, Garlic mustard, Garlicwort, Hedge garlic, Hedge mustard, Jack-by-the-hedge, Leek-cress, Penny hedge, Poor man's mustard, Poor man' treacle, Sauce Alone, Swarms.

Distribution : W. Himalaya from Kumaon to Kashmir: 2,000 – 3,500m.

Part/s Used : Whole Herb.

Utilization : Herb possesses odour resembling that of garlic, used in salads and for seasoning.

Active Principles : Herb yields a garlic scented essential oil with allyl and benzyl isothiocyanates (20-30:1), allyl sulphide and rhodanallyl.

References : Ambasta (1986), Asolkar *et al.* (1992), Bennett (1987), Chandra (1997), Kirtikar and Basu (1984).

20. *Allium ascalonicum* Linn.

Family: Liliaceae.

English and Hindi Names: **Eng.** : Eschallot, The Shellot; **Hindi** : Ekakandalasum, Gandana.

Distribution : Widely cultivated in India.

Part/s Used : Bulbs (Cloves).

Utilization : Cloves used for flavouring curries. The bulbs used for seasoning after deep frying. The flavour is similar to garlic.

Active Principles : The chief constituents identified in the oil isolated from bulbs are dimethyl disulphide, 3,4-dimethyl thiophene, dimethyl trisulphide, dipropyl disulphide, dipropyl trisulphide, methyl-cis-propenyldisulphide, methyl propyl trisulphide, 1-methyl-thiopropyl ethyl disulphide, 2- n-hexyl-5-methyl-2,3-dihydrofuran-3-one, propyl- cis-propenyl disulphide, propyl-trans-propenyl disulphide and 2-n-octyl-5-methyl-2,3- dihydrofuran-3-one. Quercitol, its 3,4'-diglucoside and 7,4'-diglucoside (I).

References : Ambasta (1986), Kirtikar and Basu (1984), Mehtani and Ingle (1999), Rastogi and Mehrotra (1990), Watt (1972).

21. *Allium carolinianum* DC.

Syn : A. blandum Wall.; *A. thomsonii* Baker.

Family : Liliaceae.

Part/s Used : Flowering Tops. Leaves. Rootstocks.

Utilization : Dried leaves and rootstocks used as spices. Leaves used in seasoning food and also for flavouring soups along with flowering tops.

References : Bennett (1987), Kapur *et al.* (1996) Negi and Pant (1990), Rawat and Pangtey (1987), Sood *et al.* (2001).

22. *Allium cepa* Linn. (Pl. 1C, 1D)

Family : Liliaceae.

English, Sanskrit and Hindi Names: **Eng. :** The Onion; **Sans.:** Dirghapatra, Mahakanda, Palandu, Rajapriya; **Hindi :** Piyaz.

Distribution : Cultivated throughout India.

Part/s Used : Bulbs (underground). Bulbils (aerial).

Utilization : Bulbs and bulbils used as condiments. Yield of volatile oil is 0.005 – 0.01%. The characteristic spicy flavour, aroma and taste of onion bulbs make it a very important flavouring ingredient of a whole range of cooking items. The oil is immensely used in the canning industry to flavour soups, meat, sauces, dressings, etc.

Active Principles : Bulbs contain diosgenin (0.18%). (+) S-methyl-L-cysteine-sulphoxide and (+) S-propyl-L-cysteine sulphoxide, propenyl sulphenic acid and lipoxygenase metabolites of linoleic acid-isomeric-9,10,13-trihydroxy-11-octadecenoic acid (43.3%). Main constituents of volatile oil are allicin (an antibiotic), allin, flavonoids, phenolic acids and sterols. Compounds contributing to the flavour profile are thiopropanal S-oxide, propyl - and propenyl containing di – and trisulphides and 2,4 - and 3,4- dimethyl thiophenes.

References : Abraham (1981), Ambasta (1986), Chevallier (1996), Harborne and Baxter (2001), Kirtikar and Basu (1984), Mehtani and Ingle (1999), Prasad and Abraham (1984), Rastogi and Mehrotra (1990, 1991, 1995), Singh *et al.* (1996), Watt (1972).

23. *Allium consanguineum* Kunth

Syn : A. stracheyi Baker

Family : Liliaceae.

Distribution : W. Himalaya: Kashmir to Kumaon.

Part/s Used : Leaves. Flowering Tops.

Utilization : Leaves used as a flavouring agent. Dried leaves, 'Jamboo' are sold in the Indian markets in large quantities. These fried with animal fat (glue) are mixed to various food preparations to increase the flavour. Also, flowering tops and leaves used as a condiment.

References : Bennett (1987), Chandra (1997), Kapur *et al.* (1996), Sood *et al.* (2001).

24. *Allium govanicum* Wall. *ex* Baker

Syn : A. humile Kunth

Family: Liliaceae.

Distribution : Temperate Himalaya: Kumaon westwards to Kashmir.

Part/s Used : Leaves. Rootstocks.

Utilization : Dried leaves known as 'Dhungar' are garlic smelling and used in seasoning the foodstuffs. Also, dried leaves and rootstocks used as spices.

References : Bennett (1987), Chandra (1997), Koelz (1997), Negi and Pant (1990), Rawat and Pangtey (1987).

25. *Allium prattii* C.H. Wright

Syn : A. victorialis var. *angustifolium* Hook.f.; *A. victorialis* non Linn.: Hook.

Family : Liliaceae.

Distribution : Sikkim.

Part/s Used : Leaves.

Utilization : Leaves used for seasoning curries.

References : Bennett (1987), Chandra (1997).

26. *Allium rubellum* M. Bieb.

Family : Liliaceae.

Hindi Names: **Hindi :** Barani-piaz, Chiri-piazi, Jangli piazi.

Distribution : Common in N.W. Himalaya: Kashmir to Kumaon.

Part/s Used : Leaves. Rootstocks.

Utilization : Sweet smelling dried leaves and rootstocks used as spices.

References : Chandra (1997), Negi and Pant (1990), Watt (1972).

27. *Allium sativum* Linn. (Pl. 1E, 1F)

Family : Liliaceae.

English, Sanskrit and Hindi Names: **Eng. :** Churl's Treacle, Garlic, Poor Man's Treacle; **Sans. :** Arishtha, Bhutabhna, Grinjana, Lashuna, Rasona, Vatari; **Hindi :** Lasan, Lahsan.

Distribution : Widely cultivated all over India.

Part/s Used : Bulbs or Cloves (underground). Bulbils (aerial).

Utilization : Bulbs consisting of cloves, used as spices and condiments for seasoning and garnishing. Bulbils also used as a condiment. Yield of volatile oil is 0.1–0.2%. Garlic oil, an important flavouring, particularly relished in meat preparations, soups, canned products, sauces, salad dressing and for flavouring vinegar.

Active Principles : Important flavour components of fresh garlic oil are allicin (2–propenyl propane thiosulphate), allin, allinase, scordinins, selenium, vitamins A, B, C and E.

References : Abraham (1981), Ambasta (1986), Bhattacharjee (1998), Chevallier (1996), Harborne and Baxter (2001), Kaul (1997), Kirtikar and Basu (1984), Mehtani and Ingle (1999), Rastogi and Mehrotra (1991, 1993, 1995), Singh and Pandey (1998), Singh *et al.* (1996), Watt (1972).

28. ***Allium schoenoprasum*** Linn.

Family : Liliaceae.

English Name : **Eng.** : The Chives or cives.

Distribution : W. Himalaya-Kashmir to Kumaon : 2,400 – 3,300m.

Part/s Used : Bulbs. Leaves.

Utilization : Onion flavoured leaves and bulbs used for seasoning salads. Leaves also used to flavour foods.

Active Principles : Bulbs yield methylpentyl disulphide and pentyl hydrodisulphide.

References : Ambasta (1986), Chandra (1997), Kirtikar and Basu (1984), Rastogi and Mehrotra (1991, 1993), Watt (1972).

29. ***Allium sphaerocephalum*** Linn.

Family: Liliaceae.

Distribution : Grows wild in Lahoul.

Part/s Used : Leaves.

Utilization : Leaves used as condiment.

References : Mittre (1981), Watt (1972).

30. ***Allium victorialis*** Linn.

Family: Liliaceae.

Distribution : Himalaya from Kashmir eastwards to Sikkim: 2,100-3,900m.

Part/s Used : Bulbs.

Utilization : Bulbs used for flavouring.

References : Asolkar *et al.* (1992), Uniyal (1968).

31. *Allium wallichii* Kunth

Family: Liliaceae.

English Name : **Eng. :** Wild Garlic.

Distribution : Temperate Himalaya (Kumaon to Sikkim): 2,400-3,900m.

Part/s Used : Leaves. Bulbs.

Utilization : Dried leaves used as condiment. Bulbs and leaves are pounded with spices to increase the flavour and aroma of food.

Active Principles : Diosgenin (0.21) and tigogenin (0.31%) isolated from bulbs.

References : Rastogi and Mehrotra (1991), Singh *et al.* (1996).

32. *Aloysia triphylla* (L'Herit.) Britton

Syn : A. citriodora Ort. *ex* Pers.; *Lippia citriodora* H. B. and K.; *Verbena triphylla* L'Herit. Stirp.

Family : Verbenaceae.

English Names : **Eng. :** Herba Louisa, Lemon Scented Verbena, Real Vervain, Sweet-scented verbena.

Distribution : Grown in Indian gardens specially in hilly regions.

Part/s Used : Leaves. Flower Tops.

Utilization : Leaves used for seasoning food. Leaves and flower tops yield an essential oil (0.07–0.7%), called Oil of Verbena with odour resembling lemongrass oil and it is used in perfumes and Eau de Cologne; and also for flavouring liquors, non-alcoholic beverages, desserts, fruit salads and jellies.

Active Principles : Salvigenin, eupatorin, eupafolin, luteolin, 6-hydroxy luteolin, luteolin-7-O- β-glucoside, hispidulin, cirsimaritin, diosmetin, chrysoeriol, apigenin, pectolinarigenin and cirsiliol isolated from leaves. The oil contains carvone, cineol, 38-40% citral, d- β-caryophyllene, iso-caryophyllene and caryophyllene-2,6- β-oxide.

References : Ambasta (1986), Asolkar *et al.* (1992), Bennett (1987), Harborne and Baxter (2001), Mehtani and Ingle (1999), Rastogi and Mehrotra (1991, 1995).

33. *Alpinia conchigera* Griff.

Family : Zingiberaceae.

Part/s Used : Rhizomes.

Utilization : Rhizomes used as a condiment.

References : Ambasta (1986), Arora (1981), Bhattacharjee (1998).

34. *Alpinia galanga* (Linn.) Willd.

Syn : A. rheedii Wight; *Amomum galanga* Lour.

Family : Zingiberaceae.

English, Sanskrit and Hindi Names : **Eng.** : The Greater Galangal, Java Galangal; **Sans.**: Aruna, Dhumala, Elaparni, Gandhamula, Kulanjan; **Hindi** : Kulanjan.

Distribution : Throughout India, often cultivated.

Part/s Used : Rhizome. Leaves.

Utilization : Rhizomes used as a spice and condiment for seasoning fish, meat and food. It yields an essential oil, used in perfumery as a source of methyl cinnamate and cineol.

Active Principles : Active principles are alpinin, galangin, kaemferide, hydroxy cinnamaldehyde, cineol and pungent oily substance galangol. Essential oil contains α- and β-pinenes, 1,8-cineole, linalool, cedrol and eugenol.

References : Ambasta (1986), Arora (1981) Bhattacharjee (1998), Islam (1996), Kirtikar and Basu (1984), Mittre (1981), Raju (2000), Rastogi and Mehrotra (1993, 1995), Saklani and Jain (1994), Sinha and Sinha (2001), Tripathi *et al.* (2001), Watt (1972).

35. *Alpinia officinarum* Hance

Family : Zingiberaceae.

English and Hindi Names: **Eng.** : The Lesser Galangal; **Hindi** : Galangal.

Distribution : South East Asia; generally cultivated.

Part/s Used : Rhizome.

Utilization : Rhizomes have a spicy odour and pungent taste resembling that of a mixture of pepper and ginger and used as flavouring ingredient in meat and meat products, condiments and relishes, curry powder blends and for flavouring liquors and tobacco. Also, it is used in perfumery preparations and nourishing spice for the digestive system.

Active Principles : Galangal rhizome yields α-pinene, cineole, a pungent oily principle galangol, linalool, methyl cinnamate, keampferide, galangin and monomethyl ether of galangin.

References : Ambasta (1986), Arora (1981), Chevallier (1996), Greenish (1999), Harborne and Baxter (2001), Rastogi and Mehrotra (1995), Watt (1972).

36. *Alpinia speciosa* K. Schum

Syn : *Alpinia nutans* Roscoe.

Family : Zingiberaceae.

English Name : **Eng.** : Light galangal.

Distribution : Cultivated in many places in India.

Part/s Used : Bulbs.

Utilization : Crushed bulbs added to food for flavouring in W. Himalaya.

References : Arora (1981), Kirtikar and Basu (1984), Mittre (1981), Watt (1972).

37. *Alpinia zerumbet* (Pers.) Burtt and Smith

Family : Zingiberaceae.

Utilization : Used as spices and condiments.

References : Arora (1981), Tripathi *et al.* (2001).

38. *Altingia excelsa* Noronha

Syn : Liquidamber altingia Bl.

Family : Altingiaceae.

Hindi Name: **Hindi :** Silaras.

Distribution : Common tree of the tropical evergreen forests in India.

Part/s Used : Timber.

Utilization : Timber yields two aromatic balsams, used in perfumery.

Active Principles : Essential oil contains benzaldehyde and cinnamic aldehyde.

References : Ambasta (1986), Asolkar *et al.* (1992), Chopra *et al.* (1956), Watt (1972).

39. *Amaranthus caudatus* Linn.

Family : Amaranthaceae.

English and Hindi Names: **Eng. :** Love lies bleeding, Tassel flower; **Hindi :** Ramdana.

Distribution : Cultivated in gardens throughout India.

Part/s Used : Leaves. Stem.

Utilization : The tender leaves and stem known for their mild flavour.

Active Principles : Leaves contain oxalic acid.

References : Asolkar *et al.* (1992), Chopra *et al.* (1956), Raju (2000), Rastogi and Mehrotra (1995), Watt (1972).

40. *Amomum aromaticum* Roxb.

Family : Zingiberaceae.

English and Hindi Names: **Eng. :** Bengal Cardamom, Jalpaiguri Cardamom; **Hindi :** Morang elaichi.

Distribution : Cultivated in the wetter districts of W. Bengal, Assam and at the foot hills of the Himalaya.

Part/s Used : Seeds.

Utilization : Seeds used as a spice.

References : Ambasta (1986), Chandra (1997), Kirtikar and Basu (1984), Tripathi *et al.* (2001), Watt (1972).

41. ***Amomum jainii*** Sunil and Ved Prakash

Family : Zingiberaceae.

Part/s Used : Rhizome.

Utilization : Rhizome used as a spice and condiment.

Reference : Tripathi *et al.* (2001).

42. ***Amomum maximum*** Roxb.

Family : Zingiberaceae.

Distribution : Introduced into W. Bengal from the Malay Islands.

Part/s Used : Fruits.

Utilization : Fruits used as spices and condiments.

References : Tripathi *et al.* (2001), Watt (1972).

43. ***Amomum subulatum*** Roxb. (Pl. 2A)

Family : Zingiberaceae.

English, Sanskrit and Hindi Names : **Eng. :** The Greater or Nepal Cardamom; **Sans. :** Aindri, Bahula, Bhadra, Bhadraila, Ela, Prithvi; **Hindi :** Bari elaichi.

Distribution : E. Himalaya. Cultivated in N. W. Bengal, Sikkim and Assam hills.

Part/s Used : Fruits. Seeds.

Utilization : Fruits used as a spice. Large Cadamom seeds with their sharp, pleasant taste resemble the true cardamom and used as a substitute for it, both as a spice and as a masticatory, Also, it forms ingredient of chewing gum and used in the preparation of sweetmeats

Active Principles : These are cineol, α-and β-pinenes, bisabolene, limonene, sabinene, terpinene, terpineol, terpinyl acetate, petunidin-3,5-diglucoside leucocyanidin-3-O- β-D-glucopyrano-side (volatile oil), cardamonin, alpinetin and subulin (seeds).

References : Anonymous (1999), Asolkar *et al.* (1992), Kirtikar and Basu (1984), Mehtani and Ingle (1999), Rastogi and Mehrotra (1991), Saklani and Jain (1994), Tripathi *et al.* (2001), Watt (1972).

44. ***Amomum vermanum*** Sunil and Ved Prakash

Family : Zingiberaceae.

Part/s Used : Fruits.

Utilization : Fruits used as spices and condiments.

References : Tripathi *et al.* (2001).

45. *Amomum xanthioides* Wall. *ex* Baker

Family : Zingiberaceae.

English and Hindi Names: **Eng. :** Malabar or Tavoy Cardamom; **Hindi :** Ilayechi – dane (the seeds).

Part/s Used : Seeds.

Utilization : Seeds posesss a strong agreeable odour.

Active Principles : Essential oil yields α-pinene, camphene, (+) limonene, linalool, camphor, (+) borneol, nerolidol and 3-phenyl-2-butanone.

References : Ambasta (1986), Rastogi and Mehrotra (1991, 1995), Watt (1972).

46. *Amphilophis odorata* A. Camus

Syn : Andropogon odoratus Lisboa

Family : Poaceae.

Distribution : W. Peninsula.

Part/s Used : Whole Herb.

Utilization : Herb smells like ginger and yields a volatile oil.

References : Ambasta (1986), Kirtikar and Basu (1984).

47. *Anadendrum montanum* Schott

Family : Araceae.

Part/s Used : Leaves.

Utilization : Leaves used for flavouring curries.

Reference : Ambasta (1986).

48. *Ananas comosus* (Linn.) Merrill

Syn : Ananas sativus Schult.f.

Family : Bromeliaceae.

English, Sanskrit and Hindi Names : **Eng. :** Ananas, Pineapple; **Sans. :** Ama, Anannasa, Kautukasanjaka, Paravati; **Hindi :** Ananas, Anannas.

Distribution : Cultivated throughout the tropics.

Part/s Used : Fruits.

Utilization : Grated or crushed fruits used in pies, ice-creams, cakes, confectionery and other preparations. The oil or essence of pineapple used for flavouring purposes in confectionery and to give the pineapple flavour to Jamaica rum.

Active Principles : Fruit contains bromelain, a protein splitting enzyme that aids in digestion, ethyl acetate, 2,3-diol diacetate vitamins A and C. The oil is a solution of othylbutyrate in alcohol.

References : Ambasta (1986), Chevallier (1996), Kirtikar and Basu (1984), Rastogi and Mehrotra (1995), Watt (1972).

49. *Anaptychia ciliaris* (L.) Koerb.

Family: Physciaceae.
Distribution : Kashmir.
Part/s Used : Thallus.
Utilization : Lichen used in perfumery.
Active Principles : Arabitol, altranorin and mannitol isolated.
Reference : Mudgal and Hajra (1997).

50. *Ancistrocladus tectorius* Merrill

Syn : Ancistrocladus extensus Wall *ex* Planch.
Family : Ancistrocladaceae.
Part/s Used : Leaves.
Utilization : Tender leaves eaten as a flavouring.
Active Principles : An alkaloid identical with ancistrocladine.
References : Ambasta (1986), Asolkar *et al.* (1992), Bennett (1987).

51. *Anethum graveolens* Linn.

Syn : Anethum sowa Kurz.; *Peucedanum graveolens* (Linn.) Benth. and Hook.f.

Family : Apiaceae.

English, Sanskrit and Hindi Names : **Eng. :** Anet, Dill, Dilly, Sowa; **Sans. :** Ahichhatra, Bahala, Chhatra, Ghosha, Misi, Poti, Shalina, Satapushpi; **Hindi :** Sowa, Soya.

Distribution : Throughout tropical and sub-tropical India. Cultivated.

Part/s Used : Leaves. Fruit. Seeds.

Utilization : Fruit and leaves used to impart flavour to curries. Dill leaves used for seasoning soups, sauces, pickles, etc. and also oil extracted from it used as soap perfume. The seeds (4%) and herb (0.5-1.2%) oil find application in compounded seasonings and flavouring for use in meat products, desserts, pickles, alcoholic beverages, etc.

Active Principles : Main constituents are carvone, coumarins, flavonoids, triterpenes and xanthones dihydrocarvone, dilla-piole, d-limonene, d-α-phellandrene, terpinene (seed oil), 3,9- epoxy-p-menth-1-ene, apiol, p-cymene, myristin, α-and β-phellandrenes and α-terpinene (Dill herb oil), carvone and d-α-phellandrene (leaves) and xanthone glycoside-dillanoside (Dill fruit). 3,6-dimethyl -2,3, 3a, 4,5, 7a - hexahydrobenzofuran contributes to the aroma of the herb.

References : Ambasta (1986), Chevallier (1996), Chopra *et al.* (1980), Greenish (1999), Kirtikar and Basu (1984), Kumar and Singh (2000), Mehtani and Ingle (1999), Rastogi and Mehrotra (1990, 1991, 1993, 1995), Watt (1972).

52. ***Anethum sowa*** Roxb. *ex* Flem.

Syn : A. graveolens DC.; *Peucedanum graveolens* C.B. Clarke

Family : Apiaceae.

English, Sanskrit and Hindi Names : **Eng. :** Indian Dil fruit; **Sans. :** Satapuspa; **Hindi :** Soya, Sova.

Distribution : Cultivated throughout India

Part / s Used : Leaves. Seeds.

Utilization : Sowa herb and oil find application as a condiment for flavouring confectionery, curries, pickles, salads and sauces. Dill seed oil (1.3-3.5%) used as a flavouring in baked goods, meat and meat products, oils and fats, condiments and relishes, alcoholic and non-alcoholic beverages and snack foods.

Active Principles : These are carvone, dihydrocarvone, dillapiole and limonene (seed oil). d–α-phellandrene, eugenol, isoeugenol, linalyl acetate, phellandral and thymol (herb oil).

References : Anonymous (1999), Harborne and Baxter (2001), Mehtani and Ingle (1999), Rastogi and Mehrotra (1991).

53. ***Angelica archangelica*** Linn. var. ***himalaica*** (C.B. Clarke) Krishna and Badhwar

Syn : Archangelica officinalis Hoffm. var. *himalaica* C.B. Clarke

Family : Apiaceae.

Distribution : Temperate Himalaya; usually near water channels in Kashmir and Sikkim.

Part / s Used : Fruits. Seeds. Leaves. Roots.

Utilization : Leaves employed as a condiment. Fruits and aromatic roots used for flavouring confectionery, liquor and wines; the latter being richer in odour and flavour. Angelica seed oil and root oil also find application as a flavour ingredient in most categories of food products, in alcoholic and non-alcoholic beverages, confectionery, baked goods, gelatins and puddings. Roots used as a seasoning agent for curry and also used in high grade perfumery.

Active Principles : Chief constituents are archangelenone (roots), umbelliprenin, isoimperatorin, bergapten, prangolarin, ostruthol, oxypeucedanin hydrate (seeds), α-pinene, β-pinene, β-phellandrene (leaves), imperatorin, isoimperatorin, bergapten and isopimpinellin (fruits). Musk-like odour of root oil has been attributed to pentadecanolide.

References : Ambasta (1986), Chevallier (1996), Harborne and Baxter (2001), Mehtani and Ingle (1999), Negi and Gaur (1994), Rastogi and Mehrotra (1990, 1991, 1993).

54. *Angelica glauca* Edgew.

Family: Apiaceae.

English Name : **Eng. :** Angelica.

Distribution : W. Himalaya: Kashmir to Shimla (2,700-3,500 m.).

Part/s Used : Roots.

Utilization : Roots added to food for flavour-like that of celery in W. Himalaya, and the essential oil extracted from it used as a substitute of European Angelica oil for use in confectioneries and preparation of liquors.

Active Principles : Roots contain angelicin, bergapten, xanthotoxin, angelicoside, archangelin and oxypeucedanin.

References : Ambasta (1986), Bhattacharjee (1998), Chandra (1997), Kapur *et al.* (1996), Mehtani and Ingle (1999), Mittre (2001), Purohit (2000), Rastogi and Mehrotra (1990, 1991, 1993, 1995), Sharma and Sood (1997), Singh (1999), Watt (1972).

55. *Anisomeles indica* (L.) Kuntze

Syn : A. ovata R. Br.

Family : Lamiaceae.

Distribution : Throughout Tropical and Sub-Tropical India, upto 1800m.

Part/s Used : Whole Plant.

Utilization : The plant has a strong camphoraceous smell, used for flavouring purposes.

Active Principles : Aerial parts rich in n-hentriacontane, glutinone, glutinol, friedelin, ovatodiolide, anisomelic acid, betulin, methyl-p-hydroxycinnamate, β-sitosterol and its glucoside.

References : Chandra (1997), Rastogi and Mehrotra (1990, 1991, 1993, 1995), Watt (1972).

56. *Annona muricata* Linn.

Family : Annonaceae.

English Name : **Eng. :** Soursop.

Distribution : Common in Tropical India.

Part/s Used : Fruits.

Utilization : Aromatic fruits used in 'sherbets' and various other drinks.

Active Principles : Fruit oil yields ethyl (E) 2-hexanoate (39.8%).

References : Ambasta (1986), Asolkar *et al.* (1992), Kirtikar and Basu (1984), Rastogi and Mehrotra (1991, 1995).

57. *Annona purpurea* Moc. and Sesse

Family : Annonaceae.

Part/s Used : Fruit pulp.

Utilization : Fruit pulp fragrant.

References : Ambasta (1986), Asolkar *et al.* (1992).

58. *Annona senegalensis* Pers.

Family : Annonaceae.

Part/s Used : Flowers.

Utilization : Flowers used in soups and as a flavouring.

References : Ambasta (1986), Asolkar *et al.* (1992).

59. *Annona squamosa* Linn. (Pl. 2B)

Family : Annonaceae.

English, Sanskrit and Hindi Names : **Eng. :** Custard apple, Sugar apple, Sweet Sop; **Sans. :** Agrimakhya, Atripya, Sitaphala, Subha, Suda; **Hindi :** At, Atasitaphal, Shariphal, Sitaphal, Sharifa.

Distribution : Cultivated throughout India.

Part/s Used : Fruits. Leaves.

Utilization : Ripe fruits delicately flavoured and made into drinks and fermented liquor. Green leaves yield pleasant-smelling essential oil (0.08%).

Active Principles : These are folic acid (fruits), polyphenols (fruit pulp), spathulenol, bornyl acetate, germacrene D, borneol and verbenone (fruit oil).

References : Ambasta (1986), Asolkar *et al.* (1992), Chopra *et al.* (1980), Chevallier (1996), Kirtikar and Basu (1984), Negi and Gaur (1994), Rastogi and Mehrotra (1995).

60. *Anthemis nobilis* Linn.

Family : Asteraceae.

English Names : **Eng. :** Double chamomile, Roman chamomile.

Distribution : Temperate regions of Asia.

Part/s Used : Flowers.

Utilization : Flowers yield an essential oil, oil of Roman or Tree Chamomile, used in perfumery and for flavouring foods, pharmaceuticals, cosmetics, alcoholic beverages, baked products and frozen dairy desserts.

Active Principles : The principle constituents of the Chamomile flowers are the volatile oil (0.8 – 1.0%) and the bitter principle, anthemic acid, and a crystalline phytosterin, anthesterin, wax, fatty oil, glucose. 3-epinobilin, 1,10-epoxynobilin and 3-dehyd-ronobilin. Flower oil yields α- and β- pinene, β-myrcene, limonene, camphene and sabinene.

References : Ambasta (1986), Asolkar *et al.* (1992), Greenish (1999), Harborne and Baxter (2001), Watt (1972).

61. ***Anthocephalus sinensis*** (Lamk.) A. Rich. *ex* Walp

Syn : A. cadamba Miq.; *A. indicus* A. Rich.

Family : Rubiaceae.

Hindi Name : **Hindi :** Kadam.

Distribution : Sub-Himalayan tracts from Nepal to Arunachal Pradesh, Bangladesh, Bhutan, Assam, Andaman and Nicobars, W. Ghats.

Part/s Used : Flowers. Fruits.

Utilization : Women decorate their coiffure with sweetly odorous 'Kadam' flowers. Fruits pleasantly flavoured.

References : Asolkar *et al.* (1992), Rastogi and Mehrotra (1993), Sahni (1998).

62. ***Anthoxanthum odoratum*** Linn.

Family : Poaceae.

English Name : **Eng. :** Sweet Vernal grass.

Distribution : Hill districts of India, either introduced or as an escape from cultivation.

Part/s Used : Whole Plant.

Utilization : Grass yields an essential oil, 'Flouve Oil' for perfume industry.

Active Principles : Coumarins are the main constituents. Its content are maximum in leaves (upto 2.5% dry basis) followed by whole plant and stems and imparts aromatic odour to the grass.

References : Ambasta (1986), Asolkar *et al.* (1992), Chopra *et al.* (1980).

63. ***Antidesma ghesaembilla*** Gaertn.

Family : Euphorbiaceae.

English and Hindi Names : **Eng. :** Black Currant Tree; **Hindi :** Umtea.

Distribution : Tropical Himalaya. E. and W. Ghats.

Part/s Used : Fruits. Leaves.

Utilization : Fruits and leaves used for flavouring meat and fish.

References : Ambasta (1986), Asolkar *et al.* (1992), Chopra *et al.* (1980), Mehtani and Ingle (1999).

64. *Apium graveolens* Linn.

Family : Apiaceae.

English, Sanskrit and Hindi Names: **Eng.** : Celery, Cultivated celery, Marsh Parsley, Smallage, Wild celery; **Sans.** : Ajamoda, Karavi, Mayura,Vishali; **Hindi** : Ajmud, Boriajmud, Karafs, Shalari.

Distribution : Foot-hills of N.W. Himalaya and outlying hills of Punjab, H.P. and Uttar Pradesh.

Part/s Used : Leaves. Fruits. Seeds.

Utilization : Blanched leaves have a typical celery odour and used to flavour dry soups, canned fish, salads and sauces. Seeds eaten as spice and also used in powder form to flavour soups, salads, tomato juice, sauces, pickles, meat preparations, fish, eggs, etc. Celery seed oil is used as such to impart a warm aromatic and agreeable flavour to food products particularly tanned soups and meat. Fruits also used as a spice and flavouring agent.

Active Principles : The flavour of celery has been attributed to a mixture of four chemicals, 3-butylphthalide, 3-butyltetrah-ydrophthalide, apiole and myristicin. The main constituents of celery seed oil are d-limonene (72.16), phthalides, β-selinene (12.17%), linalool, α-pinene, α-selinene, ligustilide, n-butyl phthalide, pentyl benzene, valerophenone, sedanoic anhydride, coumarins, apiin, and bergapten. Essential oil from ripe fruits yield more than 90% limonene, sedanolide, β-selinene, pentylcyclohexadiene and myrcene. Apigravin together with seselin, isoimperatorin, osthenol, bergapten and isopimpinellin isolated from seeds.

References : Ambasta (1986), Bhattacharjee (1998), Chevallier (1996), Harborne and Baxter (2001), Kirtikar and Basu (1984), Mehtani and Ingle (1999), Raju (2000), Rastogi and Mehrotra (1991, 1993, 1995), Watt (1972).

65. *Aquilaria agallocha* Roxb.

Syn : A. malaccensis Lamk.

Family : Thymelaeaceae.

English, Sanskrit and Hindi Names: **Eng.** : Agallochum, Aglia, Akgaw, Aloe-wood, Eagle-wood; **Sans.** : Agaru, Bhringaja, Krishna; **Hindi** : Agar.

Distribution : E. extension of Himalaya.

Part/s Used : Wood.

Utilization : The scented wood used as incense and commands a very high price. Wood on distillation yields an essential oil, 'Chuwah' or 'Agar – Attar', used in perfumery.

Active Principles : Chief ingredients are chromones-AH7,AH8 and AH9 and two tri-2-(phenylethyl) chromones-AH19a and AH19b (agar-wood), liriodenine and jinkohol (heart wood), sesquiter penoid-agarospirol (oil from plant), norsesquiterpenoids-IV,V, (S) 4a-methyl-2-(1-methylethylidene)-1,2,3,4,4a,5,6,7-octahydronaphthalene, (2R,4aS) 4a-methyl-2-(1-methyl ethenyl) 1,2,3, 4, 4a, 5, 6, 7-octahydronaphthalene, (1R, 6S, 9R) 6, 10, 10-trimethyl-11-oxatricyclo (7.2.1.0) dodecane and (1R, 2R, 6S, 9R) 6,10,10-trimethyl-11-oxatricyclom (7.2.1.0.) dodecan-2-ol (agar-wood oil).

References : Ambasta (1986), Asolkar *et al.* (1992), Bennett (1987), Kirtikar and Basu (1984), Mittre (2000), Rastogi and Mehrotra (1990, 1993, 1995), Sahni (1998), Watt (1972).

66. *Ardisia floribunda* Wall.

Syn : A. austroasiatica Walker; *A. neriifolia* Wall. *ex* DC.; *A. thyrsiflora* D.Don.

Family : Myrsinaceae.

Distribution : Sub-Tropical Himalaya from Kumaon to Bhutan; Khasia Hills upto 1,525m.

Part/s Used : Leaves.

Utilization : Dried leaves used to flavour tobacco.

References : Asolkar *et al.* (1992), Bennett (1987), Negi and Gaur (1994).

67. *Areca catechu* Linn.

Family: Arecaceae.

English Names : **Eng. :** Areca palm, Areca-nut palm, Betel-nut palm, Pinanag palm.

Distribution : Cultivated throughout tropical India; most villages in Bengal and S. India have avenues of betel palms.

Part/s Used : Nut.

Utilization : Indisponsible ingredient for the preparation of the 'pan' (betel leaf).

Active Principles : Leucocyanidin and β-sitosterol isolated from nuts.

References : Asolkar *et al.* (1992), Chopra *et al.* (1980), Rastogi and Mehrotra (1990, 1991, 1993, 1995), Watt (1972).

68. *Arecastrum romanzoffianum* Becc.

Syn : Cocos plumosa Hook. f.; *C. romanzoffiana* Cham.

Family : Arecaceae.

English Names : **Eng. :** Feathery Coconut palm, Queen palm.

Distribution : Introduced into Indian gardens.

Part/s Used : Kernels.

Utilization : Kernels yield a fatty oil similar to coconut oil in flavour.

References : Ambasta (1986), Asolkar *et al.* (1992).

69. *Armoracia rusticana* Gaertn. *et al.*

Syn : *A. lepathifolia* Gilib.; *Cochlearea armoracia* Linn.

Family : Brassicaceae.

English Name : **Eng.** : Horse – radish.

Distribution : W. Asia.

Part/s Used : Roots.

Utilization : Roots used as a condiment.

Active Principles : Roots contain glucosilinates (mainly sinigrin), asparagine, resin and vitamin C; and produce allyl isothiocyanate on crushing. Essential oil, known as horse radish oil, contains allyl isothiocyanates and allied compounds.

References : Ambasta (1986), Asolkar *et al.* (1992), Chevallier(1996).

70. *Artabotrys hexapetalus* (Linn.f.) Bhandari

Syn : *A. odoratissimus* R.Br. *ex* Ker-Gawl., non Blume; *A. uncatus* (Lour.) Baill.; *A. uncinatus* (Lam.) Merrill; *Unona hamata* Roxb.; *U. odoratissima* Roxb.

Family : Annonaceae.

Sanskrit Names: **Sans.** : Hara champaka, Nilachampaka, Phalasampenga

Distribution : Met with in W. Peninsula and cultivated throughout India.

Part/s Used : Flowers.

Utilization : Flowers yield an essential oil used in perfumery.

References : Ambasta (1986), Bennett (1987), Kirtikar and Basu (1984), Watt (1972).

71. *Artemisia absinthium* Linn.

Syn : *Absinthium officinale* Lam.; *A. vulgare* Gaertn.

Family : Asteraceae.

English, Sanskrit and Hindi Names: **Eng.** : Absinth, Madderwort, Mingwort, Mugwort, Old Woman, Warmot, Wermuth, Wormwood; **Sans.** : Damar; **Hindi** : Vilayatiafsantin.

Distribution : Kashmir: 1,700 – 2,350m.

Part/s Used : Whole Plant.

Utilization : Dried plant used as an incense. Also, yields oil isomeric with camphoraceous odour.

Active Principles : These are artemitin, rutin, absinth, artemisetin, scopoletin, umbelliferone, caffeoylquinic acid, chlorogenic acid (plant), α-thujene, α-pinene, camphene, p-cymene, 1,8-cineole, methylheptenone, β-phellandrene, caryophyllene oxide, α-terpineol, thujyl alcohol, thujyl acetate, geraniol, caryophyllene, α-himachalene, α-cadinene and elemol (essential oil).

References : Chopra *et al.* (1980), Kirtikar and Basu (1984), Rastogi and Mehrotra (1990, 1991, 1993, 1995), Sood *et al.* (2001), Watt (1972).

72. *Artemisia dracunculus* Linn.

Family : Asteraceae.

English Names : **Eng. :** Estragon, Little Dragon Mugwort, Tarragon

Distribution : Alpine regions of Kashmir and W. Himalaya: 4,200 – 4,800m.

Part/s Used : Leaves.

Utilization : Aromatic leaves used as a spice, and yield an essential oil with sweet, spicy odour, called Tarragon oil, used for flavouring canned soups, food, liquors, pickles, salad dressings, sauces and vinegar. It is also a fragrance component in soaps, detergents, cosmetics and perfumes.Yield of volatile oil is 0.8 – 1.0%.

Active Principles : The principle volatile component of Tarragon oil is methyl chavicol. Besides, elemecin, methyl eugenol, ocimene, p-methoxy-cinnamic aldehyde, sabinene and γ-terpinene are also isolated. Leaves possess scopoletin, β-sitosterol, quercetin, hyperoside and isorhamnetin.

References : Ambasta (1986), Chandra (1997), Chopra *et al.* (1960), Harborne and Baxter (2001), Kirtikar and Basu (1984), Mehtani and Ingle (1999), Raju (2000), Rastogi and Mehrotra (1991, 1995), Singh *et al.* (1996), Watt (1972).

73. *Artemisia gmelinii* Weber *ex* Stechm.

Syn : A. sacrorum Ledeb

Family : Asteraceae.

Distribution : Met with in Kumaon: 3,000 – 5,650m.

Part/s Used : Whole Plant.

Utilization : Plant yields an essential oil.

Active Principles : Essential oil contains camphor, cineol, artemisia ketone (28.0) and 1,8-cineole (13.0%).

References : Ambasta (1986), Asolkar *et al.* (1992), Bennett (1987), Kirtikar and Basu (1984), Rastogi and Mehrotra (1990, 1991, 1993, 1995), Watt (1972).

74. *Artemisia maritima* Linn.

Syn : A. fragrans Willd.

Family : Asteraceae.

English, Sanskrit and Hindi Names : **Eng.:** Drooping Sea Wormwood, English Sea Wormwood, French Sea Wormwood, Levant Sea Wormwood, Roman Wormwood, Santonica, Sea Mugwort, Sea Wormwood, Wormseed; **Sans.:** Chhara, Chauhara, Gadadhar, Gandha, Parasi, Yavani; **Hindi:** Ajavayana, Chhuhri ajmud, Kirmala, Kirmani ajwain.

Distribution : N.W. Himalaya from Kashmir to Kumaon: 2,350-6,350m.

Part/s Used : Inflorescence.

Utilization : Immature inflorescence used as an incense.

Active Principles : α-thujone and 1,8-cineole are the main active principles of which the latter attributes to sweet aroma in incen- ses and perfumes. Oil from inflorescence of mature plant conta-ins α-thujone (60.2), β-thujone (5.5) and 1,8-cineole (1.5%).

References : Chopra *et al.* (1960), Greenish (1999), Kaul (1997), Kirtikar and Basu (1984), Mittre (1981), Rastogi and Mehrotra (1990, 1991, 1995), Shah (2002), Sharma and Sood (1997), Sinha and Sinha (2001), Sood *et al.* (2001), Watt (1972).

75. *Artemisia nilagirica* (Clarke) Pamp.

Syn : A. vulgaris auct. (non Linn.); *A. vulgaris* Linn. var. *nilagirica* C.B. Clarke

Family : Asteraceae.

English, Sanskrit and Hindi Names : **Eng. :** Fleabane, Indian Wormwood, Motherwort, Mugwort; **Sans. :** Barha, Granthika, Kakapushpa, Nagadamani, Sugandha; **Hindi:** Dona, Gathivana, Majtari, Mastaru, Nagdona, Nugduna.

Distribution : Throughout the mountain tracts of India: 1,700-4,000m.

Part/s Used : Whole Plant.

Utilization : Leaf smoke used as an incense. Herb yields an essential oil used in flavouring.

Active Principles : Acetylenic compounds – trideca – 1,3,5 – trien – 7,9,11 – triyne, tetradeca -4,6 – dien – 8,10,12 – triyne – 1 – ol, heptadeca – 1,7,9 – trien – 11,13,15 – triyne, cis – dehydroma-tricarin ester and tetradeca – 6 – en – 8,10,12 – triyne – 3 – one, psilostachyin A, psilostachyin C, threonine (14.82), leucine, isoleucine (12.21), glycine (12.82), glutamic acid (9.8) and arginine (8.28%) isolated from leaves.

References : Abraham (1990), Ambasta (1986), Asolkar *et al.* (1992), Chevallier (1996), Chopra *et al.* (1980), Rastogi and Mehrotra (1990, 1991, 1995), Watt (1972).

76. *Artemisia pallens* Wall *ex* DC.

Family : Asteraceae.

Distribution : Commercially cultivated in Andhra Pradesh, Karnataka, Kerela, Maharashtra and Tamil Nadu.

Part/s Used : Whole Herb.

Utilization : Herb with its fruity fragrance is the source of Davana Oil used in high grade perfumery. The oil distilled from the flowering plants of 'Davana' also used for flavouring food items, beverages and tobacco products.

Active Principles : Herb oil contains cisdavanone, artemone, davana ether, camphene, p-cymene, sabinene, γ-terpinene, 1,8-cineole, linalool, isoborneol, geraniol, borneol, eugenol, methyleugenol, methylisoeugenol, acetoeugenol, γ-cadinene and farnesol.

References: Ambasta (1986), Asolkar *et al.* (1992), Bhakuni (2000), Bhattacharjee (1998), Kumar and Singh (2000), Rastogi and Mehrotra (1990, 1991, 1993, 1995).

77. *Artemisia roxburghiana* Bess. var. ***grata*** (Wall.) Hook.f.

Syn : Artemisia grata Wall.; *A. roxburghiana* Besser var. *roxburghii.*

Family : Asteraceae.

Part/s Used : Flowers. Leaves.

Utilization : Flowers and leaves yield an essential oil having thujone like flavour.

Active Principles : Essential oil contains β-thujone (major constituent), α-pinene, limonene, α-terpinene, 1,8-cineole, linalool, α-terpineol, β-caryophyllene, β-elemene and β-caryophullene oxide. Aerial parts yield 8-hydroxydotriacontan-27-onc together with friedelin-3-one, multiflorenone, stigmas-tanone, C14-cis-spiroketalenol ether, sesamin and β-sitosterol.

References : Ambasta (1986), Asolkar *et al.* (1992), Chopra *et al.* (1980), Rastogi and Mehrotra (1993, 1995).

78. *Artemisia stricta* Edgew.

Syn : A. edgeworthii Balak.

Family : Asteraceae.

Distribution : Himalaya: 3,000-4,500m.

Part/s Used : Leaves. Inflorescense.

Utilization : Leaves and inflorescense used as incense.

Active Principles : Oil from flower-heads contains ar-curcumene (8.21), terpinene-4-ol (7.03), camphor (4.92), cumin alcohol (4.80), cuminaldehyde (4.05), isocaryophyllene (1.30), linalool (1.03) and 3-isopropylphenol (1.03%).

References : Bennett (1987), Rastogi and Mehrotra (1995), Rawat and Pangtey (1987).

79. *Artocarpus lacucha* Ham.

Family : Moraceae.

English Name : **Eng :** Monkey Jack.

Part/s Used : Fruits.

Utilization : Dried fruits preserved as an acidulous spice for use in some curries often as a substitute for tamarind.

Active Principles : Fruits contain some water soluble bases.

References : Asolkar *et al.* (1992), Negi and Gaur (1994).

80. *Azadirachta indica* A. Juss. (Pl. 2C)

Syn : Melia azadirachta Linn.; *Melia indica* Brandis; *M. parviflora* Moon.

Family : Meliaceae.

English, Sanskrit and Hindi Names : **Eng. :** Indian lilac, Margosa tree, Neem tree; **Sans. :** Arista, Hingu, Kakaphala, Nimba, Picumarda, Prabhadra; **Hindi :** Balnimb, Nim, Nimba, Ninb.

Distribution : Cultivated all over India except Deccan peninsula.

Part/s Used : Leaves.

Utilization : Medicated soaps are prepared from Neem oil extracted from the leaves.

Active Principles : Nimbidin, azadirachtin, tiglic acid extracted from oil. Leaves contain nimocinol, 2',3'dehydrosalanno, meliacins-nimocinolide, isonimocinolide, isoazadirolide and nimbocinolide.

References : Ambasta (1986), Anonymous (1999), Bennett (1987), Bhattacharjee (1998), Chevallier (1996), Kirtikar and Basu (1984), Sahni (1998), Sinha and Sinha (2001), Watt (1972).

81. *Bauhinia variegata* Linn.

Family : Caesalpiniaceae.

English Names : **Eng. :** Buddhist Bauhinia, Mountain Ebony, Orchid Tree.

Distribution : Occuring naturally in low hills of India; cultivated throughout the plains.

Part/s Used : Flowers. Leaves.

Utilization : Flowers fragrant. Leaves used as a condiment.

Active Principles : Kaempferol glycosides isolated from flowers.

References : Asolkar *et al.* (1992), Bhattacharjee (1998), Rastogi and Mehrotra (1991), Watt (1972).

82. *Begonia jirtella* DC.

Family : Begoniaceae.

Part/s Used : Leaves.

Utilization : Used for flavouring.

Reference : Bhattacharjee (1998).

83. *Betula utilis* D.Don

Syn: *B. bhojpattra* Wall.

Family : Betulaceae.

English, Sanskrit and Hindi Names : **Eng. :** Himalayan Silver Birch, Indian Birch Tree, Indian Paper Birch; **Sans. :** Bahupata, Bhuja, Charmi, Shitri, Vidyalaya; **Hindi :** Bhujpattra, Bhejpattra, Bhujpatar.

Distribution : Temperate Himalaya: Kashmir 2,750-4,000 m, to Sikkim 3,000-4,800m.

Part/s Used : Bark.

Utilization : Bark aromatic.

Active Principles : Lupeol, betulin, oleanolic acid, karachic acid, betulic acid, acetyl oleanolic acid and leucocyanidin isolated from bark.

References : Ambasta (1986), Rastogi and Mehrotra (1990, 1991), Sahni (1998), Watt (1972).

84. *Blighia sapida* Koenig and Sims.

Syn: *Sapindus obovatus* Wight and Arn.

Family : Sapindaceae.

English Names : **Eng. :** Akee Apple, Arkee.

Distribution : Occasionally found in cultivation in India.

Part/s Used : Flowers. Aril.

Utilization : Fragrant flowers are a source of perfumed water. Parboiled aril consumed on frying; nutty in flavour and considered a delicacy.

Active Principles : Unripe fruit yields hypoglycins A and B.

References : Ambasta (1986), Asolkar *et al.* (1992), Chopra *et al.* (1980), Rastogi and Mehrotra (1990, 1991).

85. *Blumea eriantha* DC.

Family : Asteraceae.

Distribution : Bundelkhand, Konkan, Deccan, W. Ghats.

Part/s Used : Whole Herb.

Utilization : Herb yields an essential oil with camphor - like smell.

Active Principles : 5-hydroxy-3,3',4',6,7-pentamethoxyflavone isolated from herb.

References : Ambasta (1986), Asolkar *et al.* (1992), Kirtikar and Basu (1984), Rastogi and Mehrotra (1990).

86. *Blumea lacera* (Burm.f.) DC.

Syn: B. subcapitata DC.; *Conyza lacera* Burm.f.

Family : Asteraceae.

Sanskrit and Hindi Names: **Sans.:** Kukuradru, Mriduchhada, Sukshmapatra, Tamrachuda; **Hindi:** Janglimuli, Kakranda, Kakronda, Kukkurbanda.

Distribution : Throughout the plains of India from the N.W. (ascending to 700 m) to Travancore.

Part/s Used : Leaves.

Utilization : Leaves mainly yield an essential oil containing Blumea camphor.

Active Principles : These are campesterol (leaves), cineol (66%) and d-fenchone (10%) (essential oil).

References: Ambasta (1986), Asolkar *et al.* (1992), Bennett (1987), Rastogi and Mehrotra (1991), Watt (1972).

87. ***Blumea membranacea*** DC.

Family : Asteraceae.

Distribution : H.P to Sikkim. Assam. Peninsular India.

Part/s Used : Herb.

Utilization : Herb used as a flavouring agent in cooking *Dolichos lablab* fruits.

Active Principles : Herb contains sitosterol.

References : Asolkar *et al.* (1992), Kshirsagar *et al.* (2003), Rastogi and Mehrotra (1995).

88. ***Borago officinalis*** Linn.

Family : Boraginaceae.

English Name : **Eng.** : Common Borage.

Distribution : Native to Asia. Grown in gardens.

Part/s Used : Leaves. Flowers.

Utilization : Tender leaves and flowers used to impart pleasant flavour and cooling effect to beverages.

Active Principles : These are amabiline (leaves), thesinine (flowers), and choline (leaves and flowers).

References : Asolkar *et al.* (1992), Bhattacharjee (1998), Rastogi and Mehrotra (1993, 1995).

89. ***Boswellia serrata*** Roxb. *ex* Colebr.,

Syn: B. glabra Roxb.; *B. thurifera* Colebr.; *B. thurifera* Roxb. *ex* Flem.; B.S. var. *glabra* (Roxb.) Bennett; *Libanus thurifera* Colebr.

Family : Burseraceae.

English, Sanskrit and Hindi Names : **Eng.** : Indian Olibanum Tree; **Sans.** : Asraphala, Gagapriya, Hraswada, Karaka, Mocha, Rasala; **Hindi** : Kundur, Luban, Salai, Salga, Salhe, Sali, Salpe, Selgond.

Distribution : Throughout the greater part of India.

Part/s Used : Oleoresin.

Utilization : The tree yields a yellowish-green gum oleoresin with an agreeable scent when burnt.

Active Principles : Oil from oleoresin contains high content of α-thujene, serratol, 4-O-methylglucurono-arabinogalactan, 3 α-hydroxytirucall-8-24-dien-21-oic acid (I), its acetyl derivative (II), 3-ketotirucall-8,24-dien-21-oic acid (III), 3β-hydroxytirucall-8,24-dien-21-oic acid (IV), β-boswellic acid and its acetyl derivative (V), 11-keto-β-boswellic acid (VI) and its acetyl derivative (VII) digitoxose, rhamnose, xylose, arabinose, galactose and galacturonic acid.

References : Asolkar *et al.* (1992), Bennctt (1987), Jain and De (1966), Joshi (1982), Kirtikar and Basu (1984), Mittre (2001), Rastogi and Mehrotra (1991, 1993, 1995), Sahni (1998), Watt (1972).

90. *Bouea oppositifolia* (Roxb.) Meissn.

Syn: Bouea burmanica Griff.; *Mangifera oppositifolia* Roxb.

Family : Anacardiaceae.

English Name : **Eng.** : Marian Tree.

Distribution : Andaman Islands.

Part/s Used : Fruits.

Utilization : Fruits resemble small mangoes, but of inferior flavour.

References : Ambasta (1986), Bennett (1987), Watt (1972).

91. *Brassica juncea* (L.) Czernov and Coss.

Syn: B. juncea sensu Hook.f. and Anderson; *B. rugosa* var. *cuneifolia* (Roxb.); *Sinapis cuneifolia* Roxb.; *S. juncea* Linn.; *S. nurcea* Linn.; *S. ramosa* Roxb.; *S. rugosa* Roxb.

Family : Brassicaceae.

English Names : **Eng.** : Brown Mustard, Indian Mustard, The Rai.

Distribution : Cultivated abundantly in India.

Part/s Used : Seeds.

Utilization : Seeds used as condiment to flavour the curries.

Active Principles : Seeds possess sinigrin, gluconapin, sinapine, isothiocyanate (0.86%) and Me-sinapate.

References : Abraham (1981), Asolkar *et al.* (1992), Bennett (1987), Lakshamanan *et al.* (1986), Nagendra and Abraham (1984), Singh *et al.* (1996), Watt (1972).

92. *Brassica nigra* (Linn.) Koch

Syn: Sinapis erysimoides Roxb.; *S. nigra* Linn.

Family : Brassicaceae.

English, Sanskrit and Hindi Names : **Eng. :** Black Mustard, True Mustard; **Sans. :** Asuri, Jwalanti, Krimika, Madhurika, Raji; **Hindi :** Aslrai, Banarasi-rai, Ghorrai, Jagrai, Kalirai, Lahi, Makrarai, Rai, Taramira, Tira.

Distribution : Cultivated in various parts of India; chiefly on the hills.

Part/s Used : Seeds.

Utilization : Seeds have pungency and flavour and used as a condiment. Essential oil from seeds used in meat and meat products, baked goods, gelatins, puddings and non-alcoholic beverages.

Active Principles: Seeds pungent due to sinigrin, a volatile isothiocyanate (0.78%), verbenone and allyl isothiocyanate.

References : Ambasta (1986), Harborne and Baxter (2001), Kirtikar and Basu (1984), Rastogi and Mehrotra (1991, 1995), Watt (1972).

93. *Bunium penicillata* (Sesse and Moc. *ex* DC) Engl.

Syn: B. delpechiana Poiss. *ex* Engl.

Family : Apiaceae.

English Name : **Eng. :** Lignalde or Linaloe Tree.

Part/s Used : Wood. Berries.

Utilization : Essential oil from berries and wood used in perfumery, soaps and other cosmetics.

Reference : Ambasta (1986).

94. *Bupleurum candolii* Wall. *ex* DC.

Family : Apiaceae.

Part/s Used : Fruits. Roots.

Utilization : Fruits and roots used as condiments.

Reference : Negi and Gaur (1994).

95. *Bupleurum falcatum* L. var. ***manginatum*** (Wall. *ex* DC.) Cl.

Family : Apiaceae.

English Name : **Eng. :** Sickle Hare's ear.

Distribution : Mountainous tracts of N. India; Kashmir to the Khasia Hills: 1,000-4,000m.

Part/s Used : Fruits. Roots.

Utilization : Ripe fruits and dried roots used as condiments.

Active Principles : Roots contain saikosaponin a, c, d, e and f along with saikosaponins b2 and b4, angelicin, saponins-saikosides Ia, Ib and II, α-spinasteryl-β-D-glucoside.

References : Asolkar *et al.* (1992), Kirtikar and Basu (1984), Negi and Gaur (1994), Rastogi and Mehrotra (1990, 1991, 1993), Watt (1972).

96. *Bupleurum ramosissimum* Wight and Arn.

Syn: B. mucronatum var. *ramosissimum* (W.A.) Cl.; *B. wightii* var. *ramosissimum* (W. and A.) Chandrabose.

Family : Apiaceae.

Part/s Used : Whole Plant.

Utilization : Fresh plant used as a flavouring agent in curry.

References : Bennett (1987), Girach (2001).

97. *Bursera delpechiana* Poiss. *ex* Engler in DC.

Syn: B. penicillata (Sesse and Moc. *ex* DC.) Engler; *Elaphrium penicillatum* Sesse and Moc. *ex* DC.

Family : Burseraceae.

English Names : **Eng. :** Indian Lavender Tree, Lignaloe Tree.

Distribution : Introduced into India near Bangalore.

Part/s Used : Berries.

Utilization : Distillation of husk of the berries yields, 'linaloe oil', used extensively in perfumery industry.

Active Principles : Oil contains linalool and linalyl acetate.

References : Asolkar *et al.* (1992), Bennett (1987), Kumar and Singh (2000), Rastogi and Mehrotra (1990).

98. *Butea monosperma* (Lam.) Taub.

Syn: B. frondosa Koen. *ex* Roxb.; *Erythrina monosperma* Lamk.

Family : Papilionaceae.

English, Sanskrit and Hindi Names : **Eng. :** Bartard Teak, Bengal Kino, Butea Gum; **Sans.:** Bramhavriksha, Kamalasana, Palasha, Raktapushpaka, Suparni; **Hindi:** Chichra, Dhak, Kakria, Palas, Tesu.

Distribution : Common throughout the greater part of India upto 1,000 m and higher in the outer Himalaya. Hills of S. India upto 1,350 m.

Part/s Used : Root-bark.

Utilization : Root-bark added to the fermenting broth before distilling local spirits to improve flavour.

Active Principles : Bark contains Astrin.

References : Asolkar *et al.* (1992), Bennett (1987), Chopra *et al.* (1960), Kirtikar and Basu (1984), Rastogi and Mehrotra (1990), Singh and Pandey (1998).

99. *Calamintha umbrosa* Fisch and Mey

Syn : Clinopodium umbrosum Koch.

Family : Lamiaceae.

English and Sanskrit Names : **Eng. :** Catmint; **Sans. :** Karidorna.

Distribution : Temperate Himalaya. Hills of S. India.

Part/s Used : Leaves. Roots.

Utilization : Leaves used in perfumery. Roots camphoraceous.

References : Ambasta (1986), Asolkar *et al.* (1992).

100. *Callicarpa macrophylla* Vahl

Syn : C. incana Roxb., *C. roxburghii* Wall.

Family : Verbenaceae.

English and Hindi Names : **Sans. :** Phalini, Vanita; **Hindi :** Daya.

Distribution : N. and E. India, ascending the Himalaya to 2,000m.

Part/s Used : Roots.

Utilization : Roots yield an aromatic oil.

Active Principles : Glycosides, terpenes, phenolic compounds, resin, saponin isolated from roots.

References : Ambasta (1986), Anonymous (1999), Kirtikar and Basu (1984), Rastogi and Mehrotra (1995), Watt (1972).

101. *Callistemon lanceolatus* Sweet

Family : Myrtaceae.

English and Hindi Names : **Eng. :** Bottle brush; **Hindi :** Boothal brush ped.

Distribution : Grown as ornamental.

Part/s Used : Leaves.

Utilization : Leaves yield an essential oil, used in perfumery.

Active Principles : Leaves contain sitosterol, erythrodiol, betulin, betulinic, ursolic, 2 α-hydroxyursolic acids and myrtucomm-ulone A.

References : Raju (2000), Rastogi and Mehrotra (1991).

102. *Calotropis procera* (Ait.) R.Br.

Syn : C. hamiltonii Wall.

Family : Asclepiadaceae.

Distribution : Drier parts of India, chiefly in the Sub-Himalayan tract.

Part/s Used : Flowers.

Utilization : Pink flowers sweetly scented.

Active Principles : Cyanidin-3-rhamnoglucoside, quercetin-3-rutinoside, triterpene-calotropenyl acetate, multiflorenol, cyclosadol, cycloart-23-en-3β, 25-diol, β-sitost-4-en-3-one, α- and β-amyrins, stigmasterol and β-sitosterol isolated from flowers.

References : Chopra *et al.* (1960), Rastogi and Mehrotra (1991, 1995), Watt (1972).

103. *Camellia kissi* Wall.

Syn : *C. caduca* Brandis.; *C. drupifera* Dyer non Lour.

Family : Theaceae.

English Name : **Eng.** : Kissi.

Distribution : N. Bengal, E. Himalaya, Assam, Arunachal Pradesh, Khasia hills, Andaman Islands, Naga hills: 900-2,400 m.

Part/s Used : Leaves.

Utilization : The leaves mixed with tea for their pleasant fragrance.

References : Bennett (1987), Chandra (1997).

104. *Cananga odorata* (Lam.) Hook.f. and Thoms

Syn : *Canangium odoratum* Baill.; *Uvaria odorata* Lamb.

Family : Annonaceae.

English Name : **Eng.** : Ylang-Ylang Tree.

Distribution : Cultivated in many parts of India.

Part/s Used : Flowers.

Utilization : Oil from flowers, Ylang-Ylang oil used in flavouring alcoholic and non-alcoholic beverages, fruit flavours, desserts, confectionery, baked goods, gelatins and puddings. Also, used as a fragrance component in soaps, detergents, creams, lotions and perfumes, especially man's fragrances.

Active Principles : Ylang-Ylang oil contains (+)-α-pinene, linalool, geraniol, caryophyllene, cadinenes, ylangene, farnesol, farnesyl acetate, methyl benzoate, methyl salicylate, p-cresol methyl ether, benzyl acetate, safrole, isosafrole, eugenol, etc.

References : Ambasta (1986), Chevallier (1996), Harborne and Baxter (2001), Kumar and Singh (2000), Rastogi and Mehrotra (1991, 1995), Watt (1972).

105. *Canarium sikkimense* King

Syn : *C. resiniferum* Brace *ex* King; *C. strictum* Roxb.

Family : Burseraceae.

English, Sanskrit and Hindi Names : **Eng. :** Black Dammer Tree, White Dhup; **Sans. :** Mandadhupa, Raladhupa; **Hindi :** Kala Dammar.

Distribution : Evergreen forests of the W. Ghats. Arunachal Pradesh. Assam and Khasia hills.

Part/s Used : Wood.

Utilization : Wood exudes a blackish gum having a pleasant smell and burnt as incense.

Active Principles : Dammar resin contains sesquiterpene-ketone-canarone; (+) juneol, epikhusinol, ψ-taraxasterol, ψ-epitaraxasterol, α-amyrin, β-amyrin and its acetate, 11β-hydroxy-α-amyrin and 11-keto-α-amyrin.

References : Ambasta (1986), Asolkar *et al.* (1992), Bennett (1987), Kirtikar and Basu (1984), Rastogi and Mehrotra (1990), Sahni (1998), Watt (1972).

106. ***Cannabis sativa*** Linn. (Pl. 3A)

Syn : C. indica Lamk.; *C. sativa* sensu Hook.f.; C.S. var. *indica* Pers.

Family : Moraceae.

English, Sanskrit and Hindi Names : **Eng. :** Indian Hemp; **Sans. :** Ajaya, Bhanga, Chapala, Ganja, Ganjika, Jaya, Madani; **Hindi :** Bhaang, Bhanga, Charas, Ganja, Gur, Sabzi.

Distribution : Naturalized in the Sub-Himalayan tracts in India. Abundantly met with in wastelands from Punjab eastwards to Bengal and Bihar.

Part/s Used : Fruits. Seeds.

Utilization : Fruits and seeds used as a condiment.

Active Principles : Fruits contain lignanamide-cannabisins A, B, C and D. Seeds contain base-L (+)- isoleucine betaine, zeatin, zeatin nucleoside, cannabinol, 9-dehydrocannabinol, feruloylty-ramine along with p-coumaroyltyramine.

References : Anonymous (1999), Bennett (1987), Chevallier (1997), Greenish (1999), Kaul (1997), Kirtikar and Basu (1984), Rastogi and Mehrotra (1991, 1993, 1995), Shah (2002), Sharma and Sood (1997), Watt (1972).

107. ***Capparis spinosa*** (Towrn) L.

Syn : C. leucophylla DC.; *C. murrayana*, Graham; Wight.

Family : Capparidaceae.

English and Hindi Names : **Eng. :** Caper Plant; **Hindi :** Kabra, Kebra, Koyal Ber.

Distribution : C. and N. parts of the Punjab.

Part/s Used : Flower buds.

Utilization : The floral buds skin to mustard are sold as pungent condiment. Also, these are mixed in meat, pickles and sauces for producing special taste to the dishes, sauce, salads and other cooked food.

Active Principles : Flower buds contain a glycoside rutin, pectic acid, saponin, quercetin-3-rutinoside and kaempferol-3-rutinoside, kaempferol-3-O-rhamnorutinoside.

References : Bhattacharjee (1998), Chevallier (1996), Chopra *et al.* (1960), Harborne and Baxter (2001), Kaul (1997), Kirtikar and Basu (1984), Raju (2000), Rastogi and Mehrotra (1991,1993), Sharma and Sood (1997), Watt (1972).

108. ***Capsicum annuum*** Linn. var. ***acuminatum*** Fingh.

Family : Solanaceae.

English and Hindi Names : **Eng. :** Bird's Eye Chilli, Red Pepper; **Hindi :** Gachmarich, Lalmirch.

Distribution : Commonly cultivated throughout the plains of India and on the lower hills such as in Kashmir and in the Chenab valley upto 2,200m.

Part/s Used : Fruits.

Utilization : Fruits extensively used as a spice and condiment, and form the main ingredient in all chutneys and curries.

Active Principles : The fruits yield an oleoresin containing the characteristic odour and flavour of the spice, capsaicin. Besides, lutein, capsorubin, cryptoxanthin, capsanthin, rubixanthin, phytofluene and homodihydrocapsaicin are also isolated.

References : Ambasta (1986), Bennett (1987), Chevallier (1996), Kirtikar and Basu (1984), Kulkarni and Kumbhojkar (1993), Mehtani and Ingle (1999), Nagendra and Abraham (1984), Rastogi and Mehrotra (1990, 1993, 1995), Watt (1972).

109. ***Capsicum frutescens*** Linn.

Syn : C.minimum Roxb.

Family : Solanaceae.

English, Sanskrit and Hindi Names : **Eng. :** African Chillies, Bird Chilli, Cayenne Pepper, Goat Pepper, Spur Pepper Chillies; **Sans. :** Ajadakumaricha, Brahumaricha, Katuvira; **Hindi:** Gachmarich, Lalmircha, Lalmirich, Lankamirchi, Marichiphala, Ujjvala.

Distribution : Cultivated in most parts of the country, especially in Bengal, Orissa and Madras.

Part/s Used : Fruits.

Utilization : Caynne pepper much used for flavouring pickles and in hot sauces.

Active Principles : The pungent principle of capsicum fruit is due to capsaicin (0.02%).

References : Ambasta (1986), Bennett (1987), Greenish (1999), Harborne and Baxter (2001), Kirtikar and Basu (1984), Watt (1972).

110. *Cardamine pratense* Linn.

Family : Brassicaceae.

English Names : **Eng. :** Apple Pie, Cuckoo Flower, Lady Smock.

Part/s Used : Seeds.

Utilization : Seeds eaten as a condiment after meals to accelerate the digestive process.

Reference : Bhattacharjee (1998).

111. *Careya arborea* Roxb.

Family : Barringtoniaceae.

English, Sanskrit and Hindi Names : **Eng. :** Wild Guava; **Sans. :** Bhadrendrani, Girikarnika, Kalindi, Mahashwata, Nabhika, Pilu; **Hindi :** Kamba, Khumbi, Kumbh, Kumbhi, Kumbi, Kumhi, Vakamba.

Distribution : Frequent in the Sub-Himalayan tract. Bengal. S.India

Part/s Used : Flowers.

Utilization : Flowers fragrant.

References : Kirtikar and Basu (1984), Negi and Gaur (1994), Watt (1972).

112. *Carissa spinarum* Linn.

Syn : C. diffusa Roxb.

Family : Apocynaceae.

Hindi Name : **Hindi :** Karaunda.

Distribution : Throughout India especially in drier zones.

Part/s Used : Flowers.

Utilization : Flowers fragrant.

Active Principles : Berries contain coumarin.

References : Asolkar *et al.* (1992), Kirtikar and Basu (1984), Negi and Gaur (1994), Rastogi and Mehrotra (1991), Saini (1996), Watt (1972).

113. *Carum bulbocastanum* W.Koch

Syn : Bunium persicum (Boiss.) Fedts.; *C.nigrum* Royle.

Family : Apiaceae.

English and Hindi Names : **Eng. :** Black Caraway, Bulbous Caraway, Great Earth Nut, Tuberous Caraway; **Hindi :** Kala-Zirah, Shah-Zirah, Siyazira.

Distribution : Met with in Kashmir, 2,000-3,000m.

Part/s Used : Fruits. Seeds.

Utilization : Fruits and seeds used as spice for flavouring curries.

Active Principles : Essential oil (2.5%) from fruits contains p-cymol (19.15), cuminaldehyde (40.66), α- and β-pinene, limonene, camphor, acetic, propionic, butyric, oleic and benzoic acids.

References : Ambasta (1986), Asolkar *et al.* (1992), Kirtikar and Basu (1984), Rastogi and Mehrotra (1991), Sood *et al.* (2001), Watt (1972).

114. ***Carum carvi*** Linn.

Family : Apiaceae.

English, Sanskrit and Hindi Names : **Eng. :** Caraway, Common Caraway; **Sans. :** Bahugandha, Hridya, Sugandha, Sushavi; **Hindi :** Shia-Zira, Zira.

Distribution : Cultivated in the hills and plains of N.India, hills of S. India. Kashmir, Garhwal: 1,000-1,350m.

Part/s Used : Fruits. Seeds.

Utilization : Dried fruits and seeds widely used as condiments and for flavouring bread, biscuits, cakes, candies, pickles, curries, meat, sausages, vegetables, cheese, brandy (Kummelbra-nntwein) and other confectionery cordials.

Active Principles : These are lipids, umbelliferone, scopoletin, herniarin (fruits), flavonoids, p-cymene, cuminaldehyde (seed) and α - and β-pinenes, p-cymene and (+) carvone and limonene (oil) to which the odour and flavour are attributed.

References : Ambasta (1986), Asolkar *et al.* (1992), Anonymous (1999), Chevallier (1999), Chandra (1997), Greenish (1999), Harborne and Baxter (2001), Kapur *et al.* (1996), Kirtikar and Basu (1984), Mehtani and Ingle (1999), Negi and Gaur (1994), Rastogi and Mehrotra (1991, 1993, 1995), Sood *et al.* (2001), Watt (1972).

115. ***Carum copticum*** Benth. and Hook.

Syn.: Ammi copticum Boiss.; *Bunium aromaticum* Linn.; *Carum copticum* Hiern.; *Ligusticum ajawain* Fleming.; *L. ajouan* Roxb.; *Ptychotis coptica* DC.; *P. ajowan* DC.; *Sison ammi* Jacq.; *Trachyspermum ammi* (L.) Sprague; *T. copticum* Link.

Family : Apiaceae.

English, Sanskrit and Hindi Names : **Eng.:** Ajowan, Bishop's Weed, Carum, Lovage; **Sans.:** Ajamoda, Bhutika, Dipya, Hridya, Ugragandha; **Hindi :** Ajowan, Ajwain.

Distribution : Cultivated extensively in Indian gardens from the Punjab and Bengal to the S. Deccan.

Part/s Used : Fruits.

Utilization : Fruits widely used as a spice. Essential oil from fruits, 'Ajowan Oil', is aromatic.

Active Principles : Characterisctic odour attributed to thymol. The essential oil contains about 45-55% of thymol, pinenes, cymene, dipentene and carvacrol.

References : Ambasta (1986), Greenish (1999), Harborne and Baxter (2001), Kirtikar and Basu (1984), Rastogi and Mehrotra (1995), Watt (1972).

116. *Carum persicum* Boiss.

Syn : Bunium persicum (Boiss.) Fedts.; *C. bulbocastanum* C.B. Clarke in part, non W. Koch.

Family : Apiaceae.

Hindi Names : **Hindi :** Kala-Zirah, Siah-Zirah.

Distribution : Weed of high altitude cultivated fields. Found in Kashmir, Garhwal, Kumaon grasslands, Chamba, Kinnaur and Lahaul-Spiti districts of H.P.

Part/s Used : Seeds.

Utilization : Seeds used as spice.

Active Principles : These are p-cymene and cuminaldehyde (seeds), α-pinene, β-pinene, car-3-ene, limonene, γ-terpinene, p-cymene, cuminaldehyde, p-mentha-1,4-dien-7-al, p-mentha-1,3-dien-7-al, linalool, cuminyl alcohol, myristicin, carvacrol and β-phellandrene (essential oil of seeds).

References : Ambasta (1986), Asolkar *et al.* (1992), Kapur *et al.* (1996), Kaul (1997), Rastogi and Mehrotra (1991,1993), Sarin (1989), Sharma and Sood (1997).

117. *Carum roxburghianum* Benth. and Hook.f.

Syn : Apium involucratum Roxb.; *Ptychotis roxburghiana* DC.; *Trachyspermum involucratum* Wolff, non Marie; *T. roxburghia-num* (DC.) Craib.

Family : Apiaceae.

Hindi Names : **Hindi :** Ajmud, Chanu, Randhuni.

Distribution : Cultivated throughout India.

Part/s Used : Fruits. Seeds.

Utilization : Both fruits (Ajmud) and seeds used as a spice for flavouring curries, pickles, chutneys and preserves.

Active Principles: Chief constituents are 7-methoxy-6-methylcou-marin, and 7-hydroxy-6-methylcoumarin (seeds) angelicin, psoralen, bergapten, β-sitosterol, umbelliferone. d-limonene, α-terpinene, dl-piperitone (fruits).

References : Ambasta (1986), Mehtani and Ingle (1999), Rastogi and Mehrotra (1993), Watt (1972).

118. ***Caryocar nuciferum*** Linn.

Family : Caryocaraceae.

English Names: **Eng.** : Butter-Nut, Souari-Nut.

Part/s Used : Kernel.

Utilization : Kernels possess almond-like odour.

Reference : Ambasta (1986).

119. ***Caryophyllus aromaticus*** Linn. DC.

Family : Caryophyllaceae.

Distribution : Cultivated in S. India.

Part/s Used : Dried flowers/Cloves.

Utilization : The essential oil from flower buds and flower stalks of cloves have a powerful odour and flavour of cloves and much valued in perfumery. Dried flowers/cloves used to a limited extent as a hot spice throughout India.

Active Principles : Cloves contain eugenin, caryophyllin and carmufellic acid. Oil contains eugenol in variable proportions.

Reference : Watt (1972).

120. ***Cassia fistula*** Linn. (Pl. 3B)

Syn : *C. fistula* Willd.; *C. rhombifolia* Roxb.

Family : Caesalpiniaceae.

English, Sanskrit and Hindi Names : **Eng.** : Cassia, Drumstick, Golden Shower, Indian Laburnum, Pudding Pipe, Pudding Stick, Purging Cassia; **Sans.** : Amaha, Chaturangula, Dirghaphala, Rajataru, Suvarnaka; **Hindi** : Amaltas, Bandarlauri, Cirmalah.

Distribution : Lower mountainous tracts skirting the Himalaya and extending through Chota Nagpur and C. India to Bombay.

Part/s Used : Pod.

Utilization : Sweet pulp of the pods largely used in Bengal to flavour tobacco.

Active Principles : Fruit pulp contains proteins (19.94%), carbo-hydrates (26.30%), arginine, leucine, methionine, phenylalanine, tryptophan, aspartic acid and glutamic acids.

References : Anonymous (1999), Kirtikar and Basu (1984), Raju (2000), Rastogi and Mehrotra (1991, 1993), Sahni (1998), Watt (1972).

121. *Cedrela toona* Roxb. *ex* Rottl. and Willd.

Syn : Toona ciliata M.Roem.

Family : Meliaceae.

English, Sanskrit and Hindi Names : **Eng. :** Cedar, Happy Tree, Indian Mahogany, Moulmein, Sandal Neem, Singapore Cedar, Toon; **Sans. :** Nandivriksha; **Hindi :** Lim, Lud, Mahalimbu, Mahanim, Tuna, Tuni.

Distribution : Throughout the hilly districts of C. and S. India.

Part/s Used : Wood.

Utilization : Wood yields an essential oil used in perfumery.

Active Principles : Toon wood contains leucocyanidins, cedrelone copaene and mustakone. Oil contains atleast 14 components comprising essentially hydrocarbons (75%) [(-) copaene, farnesene, (-) alloaromadendrene, (+) δ-cadinene, (-) calamenene, (+) calacorene and cadalene etc.] and alcohols (25%) [(-) δ-cadinol, (-) ledol, T-muurolol, cadinol, epicubinol and cubenol.]

References : Chopra *et al.* (1980), Kirtikar and Basu (1984), Raju (2000), Rastogi and Mehrotra (1990, 1991), Watt (1972).

122. *Cedrus deodara* (Roxb.) Loud.

Family : Pinaceae.

English and Hindi Names: **Eng. :** Deodar, Himalayan Cedar; **Hindi :** Dedwar, Deodar, Deyar, Kilan, Kilan Kaper.

Distribution : Trans-Indus, N.W. Himalaya, eastwards to Kumaon; 1,200-4,000m.

Part/s Used : Heartwood.

Utilization : Essential oil extracted from scented heart-wood finds use in soap and perfumery.

References : Kirtikar and Basu (1984), Kumar and Singh (2000), Purohit (2000), Sahni (1998), Watt (1972).

123. *Cestrum parqui* L'Herit. (Pl. 3C)

Family : Solanaceae.

English Name : **Eng. :** Willow-Leaved Jessamine.

Distribution : Cultivated in Indian gardens.

Part/s Used : Whole Plant.

Utilization : Grown as an ornamental for its fragrance at night.

Active Principles : Leaves contain tigogenin, digallogenin, digitogenin and ursolic acid.

References : Ambasta (1986), Asolkar *et al.* (1992).

124. ***Chamaecyparis lawsoniana*** (Murr.) Parl.

Syn : Cupressus lawsoniana Murr.

Family : Pinaceae.

English Names: **Eng. :** Lawson Cypress, Port Oxford Cedar.

Distribution : Introduced in Indian gardens on hills as ornamental.

Part/s Used : Wood.

Utilization : Wood has a lasting spicy odour, and yields a fragrant oil.

Active Principles : These are cyanidin-3-rhamnoside and quercetin.

References : Ambasta (1986), Rastogi and Mehrotra (1990,1995).

125. ***Chenopodium ambrosiodes*** Linn.

Syn : C.valpinum Wall.

Family : Chenopodiaceae.

English Names: **Eng. :** American Wormseed, Jerusalem Oak, Mexican Tea, Sweet Pigweed.

Distribution : Many parts of India such as Bengal, Sylhet, the Deccan and Coimbatore.

Part/s Used : Fruits.

Utilization : Plant has a strong camphoraceous odour. Essential oil from fresh above ground flowering and fruiting plant, called American Wormseed Oil, used as a fragrance component in soaps, detergents and perfumes.

Active Principles : Oil contains 60-80% ascaridole, (-)-limonene, p-cymene, myrcene, α -terpinene, α -spinasterol, triacontyl alcohol and saturated hydrocarbons.

References : Chevallier (1996), Harborne and Baxter (2001), Raju (2000), Rastogi and Mehrotra (1991, 1993, 1995), Watt (1972).

126. ***Chenopodium botrys*** Linn.

Syn : C.ilicifolium Griff.

Family . Chenopodiaceae.

English Name : **Eng. :** Jerusalem Oak.

Distribution : Temperate Himalaya from Kashmir to Sikkim: 1,350-4,700m.

Part/s Used : Fruits.

Utilization : Oil used for flavouring cheese and soups.

Active Principles : α-selinene, β-selinene, β-elemene, 6 α-acetoxyselina-4(15), 11-diene, 3,11-selinadien-6 α-ol and selina-4(15), 11-diene-6α-ol and selina-4(15),11-diene-6 α-ol obtained from essential oil.

References : Asolkar *et al.* (1992), Koelz (1979), Rastogi and Mehrotra (1991, 1993), Watt (1972).

127. ***Chlorophora excelsa*** Benth. and Hook.f.

Family : Moraceae.

English Names: **Eng.** : African Oak Tree, Iroko.

Distribution : Grown in Kerela, Karnataka and Assam.

Part/s Used : Fruit Juice.

Utilization : Fruit juice widely used for flavouring foods.

Reference : Mehtani and Ingle (1999).

128. ***Chrysanthemum leucanthemum*** Linn.

Family : Asteraceae.

English Name : **Eng.** : Ox-Eye Daisy.

Distribution : Nainital, U.P.

Part/s Used : Whole Herb.

Utilization : Herb made into syrup, pastilles and essences.

Active Principles : Plant contains cyclitol, leucanthemitol and viburnitol.

References : Ambasta (1986), Asolkar *et al.* (1992).

129. ***Chrysanthemum nubigenum*** (DC.) Hand. Maz.

Family : Asteraceae.

Part/s Used : Roots.

Utilization : Roots used as an incense (Dhoop).

Reference : Negi *et al.* (1993).

130. ***Cinnamomum bejelghota*** (Hamilt.) Sweet

Syn : *C. obtusifolium* Roxb. *ex* Nees.; *Laurus bejelghota* Buch.-Ham. in Trans. Linn. Soc.; *L. cassia* in Herb. Ham.; *L. obtusifolia* Roxb.

Family : Lauraceae.

Distribution : Outer N.E. Himalaya, ascending to 2,350m. Andaman Islands.

Part/s Used : Bark. Leaves.

Utilization : Bark and leaves used for flavouring foods. Aromatic leaves used as spice.

Active Principles : Cinnamic aldehyde.

References : Ambasta (1986), Bennett (1987), Hajra and Baishya (1981), Kirtikar and Basu (1984), Mehtani and Ingle (1999), Saklani and Jain (1994), Watt (1972).

131. ***Cinnamomum burmanii*** Blume

Family : Lauraceae.

Distribution : Native to India.

Part/s Used : Bark.

Utilization : Dark bark used as a spice.

References : Ambasta (1986), Bhattacharjee (1998).

132. *Cinnamomum camphora* (Linn.) Nees and Eberm. (Pl. 4A)

Syn.: Camphora officinarum Nees. and Eberm. (L.); *Laurus camphorifera* Kamp; Roxb.

Family : Lauraceae.

English, Sanskrit and Hindi Names : **Eng. :** Camphor Cinnamon, Camphor Laurel, Camphor Tree of China, True Camphor; **Sans. :** Karpura; **Hindi :** Kapur, Karpur.

Distribution : Cultivated in India mainly in Nilgiri hills.

Part/s Used : Wood chips of stems and roots.

Utilization : Chipped wood of stems and roots is a source of natural camphor. It is used for flavouring food products particularly sweetmeats. Also, used in perfumery and other cosmetic preparations.

Active Principles : Wood contains 30-45% camphor, 5-20% cineole and up to 20% safrole. Borneol, camphene, dipentene and terpineol also present in oil.

References : Ambasta (1986), Bhattacharjee (1998), Chevallier (1996), Greenish (1999), Harborne and Baxter (2001), Kirtikar and Basu (1984), Mehtani and Ingle (1999), Vihari (1995), Watt (1972).

133. *Cinnamomum cassia* Blume

Syn : C. aromaticum Nees.

Family : Lauraceae.

English Names: **Eng. :** Cassia Bark, Cassia Lignea, Chinese Cassia, Chinese Cinnamon.

Distribution : Found in some parts of Mizoram, cultivated in Tamil Nadu.

Part/s Used : Fruits. Bark.

Utilization : Immature fruits yield an essential oil, called variously as Oil of Cassia, Oil of Cinnamon, Oleum Cinnamoni, used for flavouring processed foods and in perfuming soaps. Dried cassia buds (unripe fruits) have a faint cinnamon-like odour and a warm, sweet pungent flavour.

Active Principles : The oil from cassia buds (unripe fruits) contains cinnamldehyde as the major constituent.

References : Ambasta (1986), Greenish (1999), Harborne and Baxter (2001), Kirtikar and Basu (1984), Mehtani and Ingle (1999), Rastogi and Mehrotra (1993).

134. *Cinnamomum iners* Reinw.

Syn : Laurus nitida Roxb.

Family : Lauraceae.

Hindi Names : **Hindi :** Darchini, Jangli-Darchini.

Distribution : E. Bengal. S. India.

Part/s Used : Bark.

Utilization : Stem-bark odoriferous and used for manufacturing incense sticks (agarbattis).

References : Ambasta (1986), Kirtikar and Basu (1984), Mandal and Basu (1996), Watt (1972).

135. *Cinnamomum loureiri* Nees.

Family : Lauraceae.

English Name : **Eng. :** Saigon Cinnamon.

Distribution : S.E. Asia.

Part/s Used : Bark.

Utilization : The bark (essential oil 0.8-1.4%) used for flavouring.

Active Principles : Oil constitutes cinnamic aldehyde, phenols, pinene, phellandrene, caryophyllene.

Reference : Bhattacharjee (1998).

136. *Cinnamomum malabatrum* (Burm.f.) Blume

Syn : C. iners auct., non Reinw *ex* Blume; Hook.f.

Family : Lauraceae.

Hindi Name : **Hindi :** Jangli Darchini.

Distribution : W. Ghats. Assam. Tripura.

Part/s Used : Bark. Leaves.

Utilization : Bark and leaves used as flavouring agent.

Reference : Mehtani and Ingle (1999).

137. *Cinnamomum tamala* (F. Hamilt) Nees and Eberm (Pl. 4B)

Syn : C.albiflorum Nees. in Wall.; *Laurus cassia* Roxb.

Family : Lauraceae.

English, Sanskrit and Hindi Names: **Eng. :** Cassia Cinnamon, Indian Cassia Lignea; **Sans. :** Ankusha, Chhadana, Dala, Palasha, Tamalaka, Tejpatra; **Hindi :** Barahmi, Dalchini, Darchini, Kikoa, Tejpat.

Distribution : Tropical and Sub-Tropical Himalaya.

Part/s Used : Leaves. Bark.

Utilization : Aromatic leaves 'tejpat' added to or boiled with food- stuffs as a condiment and to flavour curries and sweets. The essential oil obtained from leaves and bark used in perfumery and in the flavouring and formulation of compounded oils for liquors and confections.

Active Principles : The oil from 'tejpat' (leaves) contains cinnamaldehyde, linalool, eugenol, eugenyl acetate, β-caryophyllene, benzaldehyde, camphor, cadinene, α-terpineol, α- and β-pinenes, p-cymene, limonene, geraniol, ocimene, γ-terpinene, β–phellandrene, benzyl cinnamate and benzyl acetate.

References : Ambasta (1986), Anonymous (1999), Bhattacharjee (1998), Dam and Hajra (1981), Jain (1968), Lalramnghinglova (2002), Mehtani and Ingle (1999), Mittre (1981), Kirtikar and Basu (1984), Purohit (2000), Rastogi and Mehrotra (1991, 1993, 1995), Sahni (1998), Saklani and Jain (1994), Singh (1999), Sinha and Sinha (2001), Watt (1972).

138. ***Cinnamomum verum*** J.S. Presl. (Pl. 4C)

Family : Lauraceae.

English, Sanskrit and Hindi Names : **Eng. :** Ceylon Cinnamon, True Cinnamon; **Sans. :** Bahugandha, Chola, Darushila, Hridya, Tamalapatra; **Hindi :** Dalchini, Darchini.

Distribution : S. India upto 1,500 m, more common even below 200 m. Also, cultivated in certain parts of India.

Part/s Used : Stem. Root-bark. Leaves.

Utilization : The aromatic stem- and root-bark employed as spices. 'Dalchini' or cinnamon spice is obtained from the inner bark of shoots in the form of rolled quills. The bark and its oil used for flavouring hot punches, spiced wines, cordials, stimulants, soups, sauces and curries, all kinds of confectionery, baked foods, meat seasonings, pickles, ketchups and table sauces, soft drinks, formulations of compound oils and in perfumery. Powdered cinnamon used in cakes, biscuits, buns and puddings. The leaf oil is a source of artificial vanilla.

Active Principles : These are α-pinene, α -phellandrene, p-cymene, linalool, β-caryophyllene, α-terpineol, benzyl acetate, cinnamic aldehyde, eugenol, cinnamyl acteate, eugenyl acetate and benzyl benzoate (leaf essential oil), and cinnamaldehyde (30.2%) (bark oil).

References: Bhattacharjee (1998), Chevallier (1996), Greenish (1999), Jain (1968), Lalramnghinglova (1996, 2002), Kirtikar and Basu (1984), Mehtani and Ingle (1999), Rastogi and Mehrotra (1993, 1995), Sahni (1998), Saklani and Jain (1994).

139. ***Cinnamomum zeylanicum*** Breyn.

Family : Lauraceae.

English, Sanskrit and Hindi Names : **Eng. :** Ceylon Cinamon; **Sans. :** Tamalapatra; **Hindi :** Dalchini.

Distribution : Cultivated in S. India.

Part/s Used : Bark. Root. Leaves.

Utilization : Root-bark yields camphor. Powdered cinnamon used in chocolates, dentifrices, incenses, perfumes and as a condiment. Cinnamon bark oil used for flavouring confectionery, liquors, pharmaceuticals and soaps whereas cinnamon leaf oil used as a flavouring agent in food, beverages, dental preparations, cosmetics and perfumes.

Active Principles : Bark oil contains 40-80% cinnamaldehyde, 4-10% eugenol, 3-5% cinnamyl acetate. Leaves contain eugenol (70-95%) followed by cinnamaldehyde (3%) and eugenol acetate.

References : Ambasta (1986), Anonymous (1999), Harborne and Baxter (2001), Watt (1972).

140. *Citrus aurantifolia* (Christm.) Swingle

Syn : *C.medica* var. *acida* Hook.f.

Family : Rutaceae.

English, Sanskrit and Hindi Names : **Eng.** : Acid Lime, Sour Lime; **Sans.** : Amlajambir, Jambira, Limpaka, Nimbuka; **Hindi** : Kaghzi Nimbu, Lebu, Limbu, Limu, Limun, Nebu, Nibu.

Distribution : Cultivated all over India.

Part/s Used : Fruits.

Utilization : Fruits and its peels extensively used for culinary purposes, flavouring jams, jellies, marmalades, alcoholic drinks, soft drinks, especially cola, lemon-lime types, ice-cream, candy, baked goods, gelatins, puddings, chewing gum, condiments, etc.

Active Principles : These are d-limonene, α- and β-pinenes, γ-terpinene, terpinolene, α-terpineol, β-bisabolene, 1,4-cineol, terpinen-4-ol and myrcene (oil), bergamotene, neral, geranial and 6,7-dihydroxy-coumarin (peels), terpenes 20.5%, alcohols 13.2%, aldehydes 36.0%, esters 23.8%, acids 2.0% and citropten 2.0% [essential oil of leaves and twigs (0.19%)].

References : Ambasta (1986), Chopra *et al.* (1980), Harborne and Baxter (2001), Kirtikar and Basu (1984), Mehtani and Ingle (1999), Rastogi and Mehrotra (1991, 1993).

141. *Citrus aurantium* Linn.

Syn : *C. aurantium* L. var. *bigaradia* Hook.f.

Family : Rutaceae.

English, Sanskrit and Hindi Names: **Eng.** : Bigarade Orange, Bitter Orange, Seville Orange, Sour Orange; **Sans.** : Brihatjambhira, Nagaranga; **Hindi** : Khatta.

Distribution : Khasia and Naga Hills.

Part/s Used : Leaves. Flowers. Fruits.

Utilization : Sour orange used for flavouring confections, marmalades, liquors and other soft drinks. Flowers, leaves and fruits yield volatile oils used in perfumery. Bitter orange peel aromatic and used to flavour food. Neroli and Petitgrain oils of bitter orange considered valuable and used for flavouring liquors and as aroma concentrates.

Active Principles : Main constituents are limonene, β-pinene, 1,8-cineol, octanal, linalool, linalyl acetate, citronellol, citronellyl formate (sour orange oil), limonene, umbelliferone, naringin, hesperidin and neohesperidin (fruits), limonene (96.41%) (peel oil), α-, β-pinenes, limonene, octanal, citronellal, linalool (11.72), linalyl acetate (18.39%), β-sitosterol (32.74), ergosterol (3.4%), desmosterol (leaf oil), linalool (50.46), limonene (9.13) and myrcene (6.26%) (flower oil).

References : Ambasta (1986), Bennett (1987), Bhattacharjee (1998), Chevallier (1996), Greenish (1999), Kirtikar and Basu (1984), Mehtani and Ingle (1999), Rastogi and Mehrotra (1991, 1993, 1995).

142. *Citrus limon* (Linn.) Burm.f. (Pl. 5A)

Syn : C. aurantium var. *limonum* W. and A.; *C. limonum* Wall.; *C. medica* L. var. *limonum* (Risso) Hook.f.; *C. medica* ssp. *limon* L.; *C. medica* Willd.; *C. pseudolimon* Tanaka.

Family : Rutaceae.

English, Sanskrit and Hindi Names : **Eng. :** Galgal, Hill Lemons, Lemon; **Sans. :** Mahanimbu, Matalunga, Matunga, Vanabija-pura; **Hindi :** Baranimbu, Jambira, Paharinimbu, Paharikaghzi.

Distribution : Cultivated in India.

Part/s Used : Flowers. Fruit peels.

Utilization : Peel fragrant are used mainly for culinary purposes and in the preparation of beverages. Lemon oil (essence) extracted from peel used for flavouring candies, pastry and baked goods, non-alcoholic beverages, ice-creams, icings, gelatins, puddings, chewing gums, cereals, meats, condiments, syrups, liquors, etc. With vanilla and other synthetic flavours, it forms the base of the flavour of custard and certain hard candies. Leaves, twigs and undeveloped fruits yield lemon petitgrain oil, used as flavourings. Essence of lemon used in perfumery and as a flavouring agent.

Active Principles : The chief constituent of lemon juice is citric acid. Oil of lemon comprises 90% of the terpene limonene and about 4-6% of aldehyde citral, the latter being the chief odorous constituent of the oil. Other odorous constituents are citronellal, geranyl acetate, linalyl acetate, octyl and nonyl aldehydes. The cold pressed lemon oil contains mainly limonene, α - and β- pinenes, γ-terpinene, sabinene, myrcene, citral and geranial. Peel contains apigenin, luteolin, chrysoseriol, quercetin, isorhamnetin, flavonols- limocitrin, limocitrol, -sinapic acid, p-coumaric acid, scopoletin, umbelliferone, hesperidin, citrusins A, B, C and D.

References : Ambasta (1986), Bennett (1987), Bhattacharjee (1998), Chevallier (1996), Chopra *et al.* (1980), Greenish (1999), Harborne and Baxter (2001), Kirtikar and Basu (1984), Mehtani and Ingle (1999), Rastogi and Mehrotra (1990, 1993, 1995), Watt (1972).

143. *Citrus maxima* (Burm.) Merrill.

Syn : Aurantium maximum Burm.; *C. decumana* L.; *C. grandis* (Linn.) Osbeck.

Family : Rutaceae.

English, Sanskrit and Hindi Names : **Eng. :** Forbidden Fruit, Grapefruit, Paradise-Apple, Pomelo, Pommel-Moose, Pompel-mos, Shaddock Pummelo; **Sans. :** Madhukarkati; **Hindi :** Batavinebu, Chakotra, Mahanibu, Sadaphal.

Distribution : Largely cultivated in Indian plains.

Part/s Used : Fruits.

Utilization : Fruits made into jams and marmalades. Also, yield an essential oil.

Active Principles : These are naringin and poncirin (peels), (+) limonene (95%) (fruit essential oil), 5-geranyloxy-7-methoxyco-umarin, isoimperatorin, bergapten and 5-geranyloxypsoralen (bergamottin) (fruits).

References : Ambasta (1986), Kirtikar and Basu (1984), Rastogi and Mehrotra (1990, 1995), Watt (1972).

144. *Citrus paradisi* Macf. (Pl. 5B)

Family : Rutaceae.

English and Hindi Names: **Eng. :** Grapefruit; **Hindi :** Chakotra.

Distribution : Cultivated in India.

Part/s Used : Rind.

Utilization : Rind yields grapefruit oil with odour intermediate between orange and lemon, employed in perfumery and for flavouring soft drinks, ice-cream, candy, baked goods, gelatins, puddings, chewing gum and toppings.

Active Principles : Grapefruit oil contains limonene, myrcene, α-pinene, sabinene, p-cymene, caryophyllene, γ-terpinene, terpinene-4-ol, octanal, linalool and nootkatone as its major components. Flavour principles are due to (+) nootkatone and 1-p-menthene-8-thiol.

References : Ambasta (1986), Harborne and Baxter (2001), Mehtani and Ingle (1999), Rastogi and Mehrotra (1991, 1995).

145. *Citrus reticulata* Blanco

Syn : C. nobilis Lour.

Family : Rutaceae.

Distribution : Widely grown in India.

Part/s Used : Fruits.

Utilization : Mandarian oil from fruits peel used for flavouring hard candies, bitter chocolates, baked goods, gelatins, puddings. chewing gum, ice-cream, non-alcoholic beverages.

Active Principles : Important constituents of Mandarin oil are limonene, γ-terpinene, p-cymene, α- and β- pinenes, sabinene, terpinolene, myrcene, linalool, α-terpincol and dimethyl anthranilate. The major flavour principles are methyl-N-methylanthranilate and thymol. Peel contains 6,7-dimethoxycoumarin, nobiletin, ponkanetin, 5-demethylno-biletin, 6,7-dimethylesculetin, 4,5,7,8-tetramethoxy-flavone, limonin, cholesterol, campesterol, stigmasterol and β-sitosterol.

References : Harborne and Baxter (2001), Mehtani and Ingle (1999), Rastogi and Mehrotra (1990, 1991, 1993, 1995).

146. *Citrus sinensis* (Linn,) Osbeck

Syn : C. aurantium var. *aurantium proper*, Hook.f.; *C. aurantium* var. *sinensis* L.

Family : Rutaceae.

Hindi Names: **Hindi :** Malta, Mosambi.

Distribution : Cultivated in India.

Part/s Used : Fruit.

Utilization : Dehydrated powdered fruits extensively used as flavouring material for cakes, pies, cookies and tobacco. The sweet orange oil generally used as a flavouring in most food products, alcoholic and non-alcoholic beverages, ice-cream, candy, confectionery, baked goods, chewing gum, gelatins and puddings, condiments, cereals, icings, meats, syrups, etc. It is mainly used as a concentrated terpeneless oil.

Active Principles : Peel contains hesperidin (0.53), narirutin (0.04%) and nobiletin. d-limonene is the main component of cold pressed orange oil, and linalool and decanal are the major oxygenated flavour compounds. Peel rich in 92-94% (+) – limonene and 2% myrcene. The characteristic orange flavour are due to octanal, neral, geranial, citronellal, α- and β- sinensals and nootkatone.

References : Bennett (1987), Harborne and Baxter (2001), Mehtani and Ingle (1999), Rastogi and Mehrotra (1990, 1991, 1995).

147. *Clausena excavata* Burm.f.

Syn : Amyris sumatrana et punctata Roxb.

Family : Rutaceae.

Distribution : Sub-Himalayan Tract. Chota Nagpur, Orissa.

Part/s Used : Leaves.

Utilization : Leaves used for flavouring curries.

Active Principles : Leaves rich in macrocyclic lactam-clausenlactam.

References : Kirtikar and Basu (1984), Mittre (1981), Rastogi and Mehrotra (1991, 1995).

148. *Clausena indica* (Dalz.) Oliver

Syn : Bergera nitida Thw., *Piptostylis indica* Dalz.; Dalz. and Gibs., Bomb.

Family : Rutaceae.

Distribution : W. Peninsula: Bombay Ghats to the Anamally hills.

Part/s Used : Leaves.

Utilization : Aromatic leaves used as condiments to flavour curries.

References : Ambasta (1986), Mehtani and Ingle (1999), Mittre (1981), Rastogi and Mehrotra (1991), Watt (1972).

149. *Clematis cadmia* Buch.-Ham.

Family : Ranunculaceae.

Part/s Used : Fruits. Roots.

Utilization : Fruits and roots used as condiments and substitute for chillies.

Reference : Chaudhuri *et al.* (1977).

150. *Cleome viscosa* L.

Syn : C. icosandra Linn.; *Polanisia icosandra* W. and A.; *P. viscosa* DC.

Family : Capparidaceae.

English, Sanskrit and Hindi Names: **Eng. :** Wild Mustard; **Sans. :** Adityabhakta, Barbara, Kukavaiminta, Raviprita, Sauri, Vikranta; **Hindi :** Hulhul, Hurhur, Jangliharrar, Kanphytia.

Distribution : Throughout the greater part of India; common in W. Bengal and S. India.

Part/s Used : Seeds.

Utilization : Seeds used for flavouring curries.

Active Principles : Seeds possess coumarinolignan-cleomiscosin A, B and C, 7-phenoxycoumarin-cleosandrin.

References : Asolkar *et al.* (1992), Kirtikar and Basu (1984), Rastogi and Mehrotra (1984), Saklani and Jain (1994), Watt (1972).

151. ***Cocos nucifera*** Linn.

Family : Arecaceae.

English, Sanskrit and Hindi Names: **Eng.** : Cocoa Nut, Cocoanut Palm, Coconut Palm, Coconut Tree; **Sans.** : Dridhaphala, Jataphala, Nadikeli, Narikela; **Hindi** : Narel, Nariel, Nariyal, Nariyalkaper, Nariyel.

Distribution : Cultivated throughout the E. and W. Coasts.

Part/s Used : Seeds.

Utilization : Seeds eaten raw or used in sweetmeats, kitchen preparation, pastries and confectionery and coconut oil, employed in food products, soaps, cosmetics, shampoos, shaving creams, etc.

Active Principles : Kernel of coconut rich in galactomannan, glycerides of lauric, myristic, palmitic, stearic, oleic, caproic, caprylic and capric acids.

References : Ambasta (1986), Chopra *et al.* (1980), Greenish (1999), Kirtikar and Basu (1984), Sahni (1998), Watt (1972).

152. ***Coffea arabica*** Linn.

Family : Rubiaceae.

English and Hindi Names : **Eng.** : Arabian Coffee; **Hindi** : Coffee.

Distribution : Cultivated in India.

Part/s Used : Roasted Beans.

Utilization : Roasted beans used for flavouring ice-creams, candies and pastries, baked goods, confectionery and carbonated beverages.

Active Principles : Main flavouring constituents are alkylpyrazines, furyl 2-methanethiol and various aliphatics. Caffeine is the main active principle.

References : Ambasta (1986), Bhattacharjee (1998), Chevallier (1996), Chopra *et al.* (1956), Harborne and Baxter (2001), Kirtikar and Basu (1984), Rastogi and Mehrotra (1990, 1991, 1995), Watt (1972).

153. ***Coffea stenophylla*** G.Don

Family : Rubiaceae.

English Name: **Eng.** : Sierra Leone Coffee.

Part/s Used : Beans.

Utilization : Beans used for flavouring ice-creams, candies and pastries and considered superior to those of *C.arabica* in flavour.

Reference : Ambasta (1986).

154. *Coleus amboinicus* Lour.

Syn : C. aromaticus Benth.; *Plectranthus aromaticus* Roxb.

Family : Lamiaceae.

English, Sanskrit and Hindi Names : **Eng. :** Country Borage, Indian Borage; **Sans. :** Pashanabhedi; **Hindi :** Pathorchur.

Distribution : Cultivated in gardens throughout India.

Part/s Used : Whole Plant specially leaves.

Utilization : The fragrant leaves used in the flavouring of meat, salad, drinks, chutneys, pickles and other culinary preparations. Also, yields an essential oil.

Active Principles : Leaves possess cirsimaritin, β-sitosterol-β-D-glucoside, oleanolic, 2 α, 3 α -dihydroxyolean-12-en-28-oic, crategolic, pomolic, euscaphic, tormentic, ursolic and 2 α, 3 α, 19 α, 23-tetrahydroxyurs-12-en-28-oic acids, salvigenin, 6-methoxy-genkwanin, quercetin, chrysoeriol, luteolin, apigenin, eriodictyol and taxifolin. Besides carvacrol (major constituent of oil responsible for the aroma), other compounds include thymol, eugenol, chavicol, ethyl salicylate, methyl eugenol, methyl chavicol, caryophyllene, cineole, d- α -terpineol and p-cymene.

References : Ambasta (1986), Bennett (1987), Kirtikar and Basu (1984), Mehtani and Ingle (1999), Rastogi and Mehrotra (1991, 1993, 1995), Watt (1972).

155. *Combretum trifoliatum* Vent.

Family : Combretaceae.

Part/s Used : Wood.

Utilization : Wood yields scented material valued in perfumery.

Reference : Ambasta (1986).

156. *Commiphora berryi* (Arn.) Engl.

Syn : Balsamodendron berryi Arn.

Family : Burseraceae.

English Name : **Eng. :** Indian Balm of Gilead.

Part/s Used : Gum-resin.

Utilization : Gum-resin used for fragrance.

References : Ambasta (1986), Bennett (1987).

157. *Commiphora caudata* (Wight and Arn.) Engl.

Syn : Protium caudatum Wight and Arn.

Family : Burseraceae.

English Name : **Eng. :** Hill Mango.

Distribution : Tamil Nadu.

Part/s Used : Gum-resin.

Utilization : Gum-resin used as incense.

References : Ambasta (1986), Asolkar *et al.* (1992), Bennett (1987).

158. ***Commiphora mukul*** (Hook. *ex* Stocks) Engl.

Syn : Balsamadendron mukul Hook. *ex* Stocks; *B. roxburghii* Stocks; *C. wightii* (Arnott) Bhandari.

Family : Burseraceae.

English, Sanskrit and Hindi Names : **Eng. :** Indian Bdellium Tree; **Sans. :** Bhutahara, Devadhupa, Guggulu, Jatala, Koushikaha, Shiva; **Hindi :** Gogil, Gugal, Guggul, Mukul.

Distribution : Met in Bellary, Mysore, Deccan, Khandesh, Kathiawar, Rajputana desert.

Part/s Used : Gum-resin.

Utilization : Finds application as a fixative in perfumery and as incense.

Active Principles : Cholesterol, 4,17 (20)-trans-pregnadien-3,16-dione, three sterols-guggusterols I, II, III, two aliphatic tetrols-octadecan-1,2,3,4-tetrol and eicosan-1,2,3,4-tetrol isolated from gum-resin.

References : Ambasta (1986), Chopra *et al.* (1980), Kirtikar and Basu (1984), Rastogi and Mehrotra (1991, 1993).

159. ***Commiphora roxburghii*** (Arn.) Engl.

Syn : Balsamodendron roxburghii Arn.

Family : Burseraceae.

Distribution : Assam, Sylhet, E. Bengal, Madhya Pradesh.

Part/s Used : Gum-resin.

Utilization : Gum-resin aromatic and used as an incense and fixative in perfumery.

References : Ambasta (1986), Chopra *et al.* (1956).

160. ***Cordia dichotoma*** Forester.f.

Syn : C. indica Lamk.; *C. latifolia* Roxb.; *C. lowriana* Brandis; *C. myxa* Roxb. non Linn.; *C. myxa sensu* Cl.; *C. obliqua* Willd.

Family : Boraginaceae.

Distribution : Common throughout India; often planted.

Part/s Used : Flowers.

Utilization : Flowers fragrant.

References : Bennett (1987), Chopra *et al.* (1956), Negi and Gaur (1994), Rastogi and Mehrotra (1991).

161. *Coriadrum sativum* Linn. (Pl. 6A)

Family : Apiaceae.

English, Sanskrit and Hindi Names: **Eng. :** Coriander Fruit; **Sans. :** Allaka, Dhanya, Janapriya, Kustumburi, Nisara; **Hindi :** Dhania, Dhaniya.

Distribution : Cultivated all over India.

Part/s Used : Leaves. Fruits. Seeds.

Utilization : Coriander leaves used for garnishing culinary preparations and also in salad, soups and chutney. Essential oil from fruits, called coriander oil used for flavouring gin, and in cocoa and chocolate industries. The powdered fruit used as a flavour to mask the taste of active purgatives. Coriander seeds, extensively used as a chief ingredient of curry powder, pickles, spices, sauces, etc. An oleoresin obtained from the seeds also finds application in flavouring beverages, pickles, sweets, etc. Besides, coriander enters into the manufacture of imitation flavours, flavourings for baking products, meat, fish, salads, soda and syrup, gelatin, dessert, pudding, candy, chocolate and other confectionery, and to flavour tobacco.

Active Principles : Dodecanal and 7-dodecenal are the most important aroma constituents in coriander leaf. Fruit yields 0.3-2.0% oil, of which 60-80% is (+) linalool, 5% camphor, 5% γ-terpinene, 5% α-pinene, mono and poly unsaturated fatty aldehydes, gnaphaloside A, gnaphaloside B, quercetin, isorhamnetin, rutin and luteolin. Seeds rich in palmitic, petrosalinic, oleic, linoleic, lauric, myristic, myristoleic and palmitoleic acids. Coriander seed oil rich in δ-linalool, linalyl acetate, thymol, β-caryophyllene, α-pinene, terpinene, borneol, limonene, β-phellandrene, citronellol, 1,8-cineole and geranyl acetate.

References : Abraham (1981), Ambasta (1986), Anonymous (1999), Bhattachrjee (1998), Chevallier (1997), Chopra *et al.* (1980), Greenish (1999), Hajra and Baishya (1981), Harborne and Baxter (2001), Kirtikar and Basu (1984), Mehtani and Ingle (1999), Rastogi and Mehrotra (1990, 1991, 1993, 1995), Singh *et al.* (1996), Watt (1972).

162. *Crocus sativus* Linn. (Pl. 6B)

Family : Iridaceae.

English, Sanskrit and Hindi Names: **Eng. :** Saffron, Saffron Crocus, Spanish Saffron; **Sans. :** Agnishikha, Balhika, Charu, Dhira, Kaisara; **Hindi :** Kesar, Zafran.

Distribution : Cultivated in Jammu and Kashmir.

Part/s Used : Dried stigmas. Tops of the styles.

Utilization : Stigmas and tops of styles used as a flavouring agent for imparting pleasing aroma to foods, cottage cheese, bouillabaisse, mayonnaise, rice, cakes, sweetmeats, chicken, meat, liquors and cordials. A tincture of saffron oil used to flavour essences.

Active Principles : Stigma and style possess crocins. Its volatile oil contains several cyclohexanone derivatives including safranal.

References : Ambasta (1986), Greenish (1999), Kaul (1997), Kirtikar and Basu (1984), Mehtani and Ingle (1999), Misra and Kumar (2000), Watt (1972).

163. *Crypteronia paniculata* Blume

Syn : C. glabra (Wall.) Planch. *ex* Endl.; *C. pubescens* Blume.

Family : Sonneratiaceae.

Part/s Used : Shoots.

Utilization : Young shoots used for flavouring.

References : Ambasta (1986), Bennett (1987), Watt (1972).

164. *Cryptomeria japonica* (Linn.f.) D.Don

Family : Pinaceae.

English Name : **Eng. :** Japanese Cedar.

Distribution : Cultivated throughout the districts of Darjeeling, Shimla and occasionally in other hill stations.

Part/s Used : Leaves.

Utilization : Leaves used as incense sticks.

Active Principles : Leaf oil contains 4-terpineol, monoterpene hydrocarbons (45.0), monoterpene alcohols (10.0), monoterpene esters (2.0), sesquiterpenes (25.0) and diterpenes (12.0%).

References : Ambasta (1986), Rastogi and Mehrotra (1995), Watt (1972).

165. *Cucumis sativus* Linn.

Syn : C. hardwickii Royle; *C.s.* var. *sikkimensis* Hook.f.

Family : Cucurbitaceae.

English, Sanskrit and Hindi Names: **Eng. :** Cucumber; **Sans. :** Bahuphala, Kandalu, Pitapushpa, Trapushpa; **Hindi :** Khira.

Distribution : Cultivated in all parts of India.

Part/s Used : Fruits. Seeds.

Utilization: Odorous cucumber used in blending certain bouquet perfumes. Seeds also used in perfumery.

Active Principles : Fruit contains rutin and volatile compounds whereas seeds rich in cucurbitaside.

References : Ambasta (1986), Asolkar *et al.* (1992), Bennett (1987), Kirtikar and Basu (1984), Rastogi and Mehrotra (1991, 1993, 1995), Vihari (1995), Watt (1972).

166. *Cuminum cyminum* Linn. (Pl. 6C)

Family : Apiaceae.

English, Sanskrit and Hindi Names: **Eng.** : Cumin, Cumin Seed; **Sans.** : Jira, Jiraka; **Hindi** : Jira, Zeera.

Distribution : Cultivated in most parts of India, except Bengal and Assam.

Part/s Used : Fruits.

Utilization : Fruits employed as a spice and condiment in curry powder to flavour dishes. It yields, oil of cumin which finds application in flavouring curries and other culinary prepara-tions, confectionery, alcoholic and non-alcoholic beverages, liquors, cordials; also used as a fragrance component in creams, lotions and perfumes.

Active Principles : The characteristic odour of cumin oil attributed to cuminaldehyde, 1,3-p-menthadien-7-al and 1,4-p-menthadien-7-al. Volatile oil yields 25-35% aldehydes, pinene and α-terpineol.

References : Ambasta (1986), Asolkar *et al.* (1992), Anonymous (1999), Chevallier (1996), Greenish (1999), Harborne and Baxter (2001), Mehtani and Ingle (1999), Nagendra and Abraham (1984), Rastogi and Mehrotra (1991, 1995), Singh and Pandey (1998), Watt (1972).

167. *Cupressus lusitanica* Mill.

Family : Cupressaceae.

English Names : **Eng.** : Goa Cypress, Mexican Cypress.

Distribution : Cultivated in W. India above the Ghats and Goa.

Part/s Used : Leaves.

Utilization : Leaves yield an aromatic oil.

Active Principles : Oil possesses monoterpene hydrocarbons (66.0%) and α-pinene (18%).

References : Ambasta (1986), Rastogi and Mehrotra (1991,1995), Watt (1972).

168. *Cupressus sempervirens* Linn.

Family : Cupressaceae.

English, Sanskrit and Hindi Names: **Eng.** : Mediterranean Cypress; **Sans.** : Surahva; **Hindi** : Sara, Saras, Saru.

Distribution : N.W. India.

Part/s Used : Needles.

Utilization : Oil of Cypress extracted from needles used as a fragrance component in colognes, aftershaves and perfumes.

Active Principles : The oil mainly contains α-pinene, car-3-ene, camphene, sylvestrene, cymene and sabinol.

References : Ambasta (1986), Chevallier (1996), Harborne and Baxter (2001), Kirtikar and Basu (1984), Rastogi and Mehrotra (1990, 1991, 1993, 1995), Watt (1972).

169. *Curcuma amada* Roxb.

Family : Zingiberaceae.

English Name : **Eng. :** Mango-Ginger.

Distribution : Bengal and on the hills.

Part/s Used : Rhizome.

Utilization : Rhizome used as a condiment and it is akin to fresh mangoes. Also, used in perfumery.

Active Principles : Rhizomes yield d-α-pinene, d-camphor, 1-β-curcumene as hydrochloride, d-curcumene and phytosterol.

References : Chevallier (1996), Rastogi and Mehrotra (1990), Watt (1972).

170. *Curcuma angustifolia* Roxb.

Family : Zingiberaceae.

English, Sanskrit and Hindi Names : **Eng. :** East Indian Arrowroot, Narrow Leaved Turmeric, Wild Arrowroot; **Sans. :** Gavayod-bhava, Pishtika, Tavakshir; **Hindi :** Tavakhira, Tokhur.

Distribution: Outer ranges of C. Himalaya, W.Bihar, N.Bengal, extending to Bombay and S. India.

Part/s Used : Roots. Tubers.

Utilization : Roots fragrant. Tubers used as condiments.

References : Arora (1981),Tripathi *et al.* (2001), Watt (1972).

171. *Curcuma aromatica* Salisb.

Syn : C. zedoaria Roxb.

Family : Zingiberaceae.

English, Sanskrit and Hindi Names : **Eng. :** Cochin Turmeric, Wild Turmeric, Yellow Zedoary; **Sans. :** Aranyaharidra, Sholi, Vanharidra; **Hindi :** Ban-haldi, Banharidra, Janglihaldi.

Distribution : Bengal, W. Peninsula; sometimes cultivated.

Part/s Used : Rhizomes.

Utilization : Rhizomes yield an essential oil.

Active Principles : Essential oil contains xanthorrhizol (25.7), β-curcumene (25.5) and ar-curcumene (18.6%).

References : Arora (1981), Kirtikar and Basu (1984), Rastogi and Mehrotra (1993, 1995), Tripathi *et al.* (2001), Watt (1972).

172. ***Curcuma caesia*** Roxb.

Family : Zingiberaceae.

English and Hindi Names : **Eng. :** Black Zedoary; **Hindi :** Nar-Kachura.

Distribution : Common in gardens in Bengal.

Part/s Used : Rhizome.

Utilization : Aromatic rhizomes used in cosmetic. The oil from rhizome considered a good source of camphor.

Active Principles : These are (+) linalool (20.42), (+) camphor (18.88), ocimene (15.66), 1-ar-curcumene (14.84), zingiberol (12.60), 1,8-cineole (9.06) and (+) borneol (7.0%) (rhizome essential oil).

References : Ambasta (1986), Arora (1981), Rastogi and Mehrotra (1995), Tripathi *et al.* (2001), Watt (1972).

173. ***Curcuma domestica*** (Medik.) Valeton (Pl. 6D, 6E)

Syn : *C. longa* Linn.

Family : Zingiberaceae.

English, Sanskrit and Hindi Names: **Eng. :** Indian Saffron, Turmeric; **Sans. :** Aneshta, Bahula, Haridra, Gauri, Nisa, Rajani, Ralri; **Hindi :** Halada, Haldi, Hardi.

Distribution : Mainly cultivated in S. and E. India.

Part/s Used : Rhizomes.

Utilization : The flavourant rhizomes used as a condiment. Also, it yields an oil.

Active Principles : Chief constituents of turmeric oil responsible for the flavour are ar-, α- and β- turmerones, zingiberene. α- and β-pinenes, camphene, limonene, terpinene, caryophyllene, curcu-mene, linalool, turmerone, ar-turmerone, borneol, isoborneol, camphor, eugenol, cineole, curdione, curzerenone, curcumin, monodemethoxy-curcumin and didemethoxy-cucurcumin.

References : Ambasta (1986), Anonymous (1999), Arora (1981), Bennett (1987), Bhattacharjee (1998), Chaudhari and Pal (1981), Francis (1984), Chevallier (1996), Greenish (1999), Mehtani and Ingle (1999), Kirtikar and Basu (1984), Kulkarni and Kumbhojkar (1993), Rastogi and Mehrotra (1991, 1993, 1995), Tripathi *et al.* (2001).

174. ***Curcuma zedoaria*** Rosc.

Syn : *C. zerumbet* Roxb.

Family : Zingiberaceae.

English, Sanskrit and Hindi Names : **Eng. :** Long Zedoary, Round Zedoary; **Sans. :** Dravida, Gandhasara, Jatala, Kalpaka, Mukhya, Vedhya, Sati; **Hindi :** Kachura, Kalihaldi.

Distribution : Wild in the E. Himalaya. Cultivated more or less throughout India.

Part/s Used : Rhizome.

Utilization : Fragrant rhizome finds use as condiment and spices and in native perfumery. Also, it yields an essential oil.

Active Principles : These are curzerenone (essential oil), sesquiterpenoids - curcumol, pyrocurzerenone, guaiane-zedoaro-ndiol, more sesquiterpenoids-(-) curcumenone, curcumanolides A and B, zedoarol, 13-hydroxy germacrone, curzeone, furanogermenone and curcumenol (rhizomes).

References : Ambasta (1986), Arora (1981), Chevallier (1996), Kirtikar and Basu (1984), Rastogi and Mehrotra (1990, 1991, 1995), Tripathi *et al.* (2001), Watt (1972).

175. ***Cydonia oblonga*** Mill.

Syn : C. vulgaris Pers.; *Pyrus cydonia* Linn.

Family : Rosaceae.

English, Sanskrit and Hindi Names : **Eng. :** Quince; **Sans. :** Amritphala; **Hindi :** Bihi.

Distribution : Cultivated in Punjab, Kashmir and the Nilgiris.

Part/s Used : Fruit.

Utilization : Fruits fragrant and used as a flavouring to cooked apples.

Active Principles : The flavour principle of fruit is ethyl 2-methyl-2-butanoate.

References : Asolkar *et al.* (1992), Bennett (1987), Chevallier (1996), Chopra *et al.* (1956), Harborne and Baxter (2001), Raju (2000), Rastogi and Mehrotra (1991, 1993, 1995).

176. ***Cymbopogon caesius*** Stapf

Syn : Andropogon caesius Nees.; *A. schoenanthus* Linn. var. *caesius* Hack.

Family : Poaceae.

English Name : **Eng. :** Kachi Grass.

Distribution : Neighbourhood of Bangalore.

Part/s Used : Grass.

Utilization : Grass yields an essential oil, Kachi grass oil, resembling ginger grass oil in odour and properties.

References : Ambasta (1986), Asolkar *et al.* (1992).

177. ***Cymbopogon citratus*** (DC.) Stapf

Syn : Andropogon citratus DC.

Family : Poaceae.

English, Sanskrit and Hindi Names: **Eng. :** Melissa Grass, West Indian Lemon Grass; **Sans. :** Atigandha, Badhira, Bhustrina, Chhatra; **Hindi :** Agin Ghas, Gandhatrina.

Distribution : Cultivated in India.

Part/s Used : Grass.

Utilization : Lemon grass oil extracted from grass finds use in flavouring foods, alcoholic and non-alcoholic beverages, baked goods, meat and meat products, oils and fats, confectionery and desserts, cosmetics, toiletries, detergents and perfumes.

Active Principles : Essential oil contains α- and β- citrals (72.29), myrcene (1.5), methylheptenone (0.2), linalool (1.2), linalyl acetate (0.1), 2-undecanone (0.3), geranyl acetate (0.1), citronellol (1.5) and nerol (0.8%).

References : Ambasta (1986), Chevallier (1996), Harborne and Baxter (2001), Jain (1968), Jain and Sharma (2000), Kirtikar and Basu (1984), Raju (2000), Rastogi and Mehrotra (1991, 1993, 1995), Sharma and Sood (1997), Watt (1972).

178. ***Cymbopogon coloratus*** (Nees *ex* Hook.f.)

Syn : Andropogon coloratus Nees.; *A. nardus* Linn. var. *coloratus* Hook.f.

Family : Poaceae.

Distribution : C. to Peninsular India.

Part/s Used : Grass.

Utilization : Oil extracted used for perfuming soaps.

Active Principles : Grass oil comprises of geraniol (69.11), limonene (5.81), geranyl acetate (3.80), trans-β-ocimene (3.46), neral (2.27), elemol (2.04), myrcene (1.76), linalool (1.69) and geraniol (1.39%).

References : Ambasta (1986), Rastogi and Mehrotra (1995).

179. ***Cymbopogon flexuosus*** (Nees *ex* Steud.) Wats.

Syn : Andropogon nardus L.var. *flexuosus* Hack.

Family : Poaceae.

English Name : **Eng. :** Cochin Lemon Grass, East Indian Lemon Grass, Malabar Lemon Grass.

Distribution : S. India. U.P. Sikkim.

Part/s Used : Leaves.

Utilization : Extracted oil finds use in perfumery, soaps and cosmetics.

Active Principles : Oil mainly contains α-terpinene, myrcene, ocimene, dipentene, α-pinene, p-cymene and citral.

References : Ambasta (1986), Bhakuni (2000) Jain (1968), Kumar and Singh (2000), Rastogi and Mehrotra (1991, 1993, 1995).

180. *Cymbopogon martini* (Roxb.) Wats.

Syn : *Andropogon martini* Roxb.; *C. motia* Gupta; *C. schoenanthus* Linn. var. *martini* Hook.f.

Family : Poaceae.

English, Sanskrit and Hindi Names : **Eng. :** Rosha Grass, Rusa Grass; **Sans. :** Rohisa; **Hindi :** Gandh Bel.

Distribution : Drier localities of Kashmir, Punjab, Kumaon, Garhwal, Rajasthan, Maharashtra, Madhya Pradesh, S. India.

Part/s Used : Leaves. Flowering Tops.

Utilization : Leaves added to kitchen preparations to flavour them. Extracted oil, variously called as Palmarosa Oil, Rusa Oil, or East Indian geranium oil finds extensive use in soaps and cosmetics. Also, used extensively in India as an adulterant of 'Attar of Roses'.

Active Principles : Chief constituents of essential oil are geranyl acetate, β-caryophyllene, caryophyllene epoxide, linalool, geraniol and geranyl n-butyrate.

References : Ambasta (1986), Bennett (1987), Bhakuni (2000), Kumar and Singh (2000), Jain (1968), Raju (2000), Rastogi and Mehrotra (1993, 1995), Singh and Pandey (1998).

181. *Cymbopogon nardus* (Linn.) Rendle

Syn : Andropogon nardus Linn.

Family : Poaceae.

English and Hindi Names: **Eng. :** Citronella Grass; **Hindi:** Ganjnika-ghas, Pustbura.

Distribution : Throughout the hotter parts of India.

Part/s Used : Grass.

Utilization : Leaves used in the flavouring of foods, baked goods, alcoholic and non-alcoholic beverages, confectionery and frozen dairy desserts. Essential oil extracted from leaves, citronella oil or Ceylon citronella oil, finds use in perfumery for scenting soaps, varnishes, insecticides, spraying liquids and shoe polishes.

Active Principles : Leaves contain 0.5-1.2% of oil, comprising 25-55% (+)-citronellal, 10-15% (+)-citronellol, 20-40% geraniol and 7-10% geranyl acetate.

References : Ambasta (1986), Jain (1968), Harborne and Baxter (2001), Kirtikar and Basu (1984), Rastogi and Mehrotra (1990, 1991, 1995).

182. ***Cymbopogon pendulus*** (Nees *ex* Steud.) Wats.

Syn : Andropogon pendulus Nees *ex* Steud.

Family : Poaceae.

Distribution : N.E. India.

Utilization : Plant aromatic and employed for the production of lemongrass oil.

Active Principles : Essential oil contains 78.5% citral.

References : Asolkar *et al.* (1992), Bhakuni (2000), Kumar and Singh (2000), Rastogi and Mehrotra (1991, 1995).

183. ***Cymbopogon winterianus*** Jowitt

Family : Poaceae.

English Name : **Eng. :** Winter Grass.

Distribution : Grown in N. E. States and S. India.

Part/s Used : Leaves.

Utilization : Much valued for its citronella java oil (obtained from leaves), used in perfumery.

Active Principles : Essential oil yields citronellal, geranyl acetate, geraniol, citronellol, isopulegol, eugenol, elemol, citronellic acid and hydroxycitronellic acid.

References : Kumar and Singh (2000), Rastogi and Mehrotra (1993, 1995).

184. ***Cyperus articulatus*** Linn.

Family : Cyperaceae

English Name : **Eng. :** Guinea Rush.

Distribution : Warm regions.

Part/s Used : Tubers.

Utilization : Fragrant tubers used for perfuming clothing.

References : Ambasta (1986), Asolkar *et al.* (1992), Kirtikar and Basu (1984).

185. ***Cyperus bulbosus*** Vahl

Family : Cyperaceae.

Distribution : Found throughout India.

Part/s Used : Tubers.

Utilization : Roasted tubers baked into bread or cooked into puddings for the flavour of potatoes.

Active Principles : Tubers contain caryophyllene oxide and humulene oxide.

References : Ambasta (1986), Rastogi and Mehrotra (1995).

186. ***Cyperus kyllingia*** Endl.

Syn : C. monocephalus (Rottb.) F.V.M., Fragm., non Roxb.; *Kyllingia monocephala* Rottb.

Family : Cyperaceae.

Sanskrit and Hindi Names : **Sans. :** Musta, Nirvisha; **Hindi :** Nirbishi, Shwetgothubi.

Distribution : Peninsular India.

Part/s Used : Rhizome.

Utilization : Aromatic rhizomes yield an essential oil.

References : Ambasta (1986), Bennett (1987), Watt (1972).

187. ***Cyperus metzii*** (Hochst.) Mattf. and Kukenthal

Syn : Kyllingia squamata Vahl.

Family : Cyperaceae.

Distribution : W. Himalaya. W. Ghats.

Part/s Used : Rhizome.

Utilization : Juice of the rhizome used as flavouring agent for foods and medicine.

References : Ambasta (1986), Mehtani and Ingle (1999).

188. ***Cyperus rotundus*** Linn.

Syn : C. hexastachyos Roxb.; *C. retzii* Nees. in W. Contrib., non Poir.; *C. tuberosus sensu* Cl. non Rottb.

Family : Cyperaceae.

English, Sanskrit and Hindi Names : **Eng. :** Nut Grass; **Sans. :** Abda, Bhadrakshi, Musta, Mustaka; **Hindi :** Motha, Mutha.

Distribution : Kuram Valley. Gilgit. Kashmir to Shimla. Garhwal. Khasia Hills.

Part/s Used : Tuberous Roots. Rhizomes.

Utilization : Roots and rhizomes used for making fragrant screens. Dried tuberous roots, Soucher, employed in low grade perfumes and Agarbattis. Tubers used as a flavouring agent and in perfumery. Rhizomes also yield an essential oil used in dyeing to impart a perfume to the fabric.

Active Principles : These are 4 α, 5 α-oxidoeudesm-11-en-3 α-ol (I), cyperene, β-selinene, cyperenone, α-cyperone. Tubers contain cyperene-1 (a tricyclic sesquiterpene) and cyperene-2 (a bicyclic sesquiterpenen hydrocarbon), cyperotundone, and oleanolic acid-3-O-neohesperidoside (I) (rhizome), sesquiterpene-ketone-mustakona (essential oil).

References : Ambasta (1986), Bennett (1987), Kirtikar and Basu (1984), Kshirsagar *et al.* (2003), Raju (2000), Rastogi and Mehrotra (1990, 1991, 1993, 1995), Singh and Pandey (1998), Watt (1972).

189. *Cyperus scariosus* R.Br.

Syn : *C. pertenuis* Roxb.

Family : Cyperaceae.

Sanskrit and Hindi Names : **Sans. :** Chakranksha, Kalapini, Nagarotha, Nargmusta; **Hindi :** Nagarmutha.

Distribution : Damp places in Bengal. Rare in Punjab.

Part/s Used : Tubers.

Utilization : Aromatic tubers used in perfumes and scented Agarbattis. Essential oil obtained considered a substrate for Patchouli, used as a hair perfume and also in dyeing to give a scent to the fabric.

Active Principles : Oil contains rotundene, rotundenol, (-) β-selinene, cyperenone, cyperenol and patchoulenol.

References : Ambasta (1986), Kirtikar and Basu (1984), Rastogi and Mehrotra (1990, 1991, 1995), Watt (1972).

190. *Daphne retusa* Hemsl.

Family : Thymelaeaceae.

Utilization : Inflorescence burnt as incense.

Reference : Rawat and Pangtey (1987).

191. *Daucus hypoleuca* Wedd.

Family : Apiaceae.

Part/s Used : Fruits.

Utilization : Fruits used as a flavouring.

Reference : Ambasta (1986).

192. *Daucus sativa* DC.

Syn : Daucus carota Linn.; *D. carota* L. ssp. *carota*

Family : Apiaceae.

English, Sanskrit and Hindi Names : **Eng. :** Bee's nest, Bird's nest, Cultivated Carrot, Dauke; **Sans. :** Dindiramodaka, Gajara, Kanda, Naranga, Shikha-mula; **Hindi :** Gagar, Gajar, Gajra.

Distribution : Kashmir and W. Himalaya: 1,700-3,000m.

Part/s Used : Roots. Seeds.

Utilization : Carrot seed oil much valued in spice blends, formulation of compound oil for sauces, cured meats, seasonings, flavouring alcoholic and non-alcoholic beverages, baked goods, confectionery, condiments, gelatins, puddings, and also as a fragrance component in soaps, detergents, perfumes and lotions.

Active Principles : Important constituents are carotol (18-50%), α-pinene (10-13%), β-pinene, geranyl acetate (up to 10.5%), daucol, limonene, geraniol, β-bisabolene, caryophyllene, β-elemene, eugenol, asarone (seed essential oil). α-humulene, β-farnesene and α-bergamotene (volatile oil of roots) and umbelliferone (roots).

References : Ambasta (1986), Chevallier (1996), Harborne and Baxter (2001), Mehtani and Ingle (1999), Rastogi and Mehrotra (1990, 1991, 1993, 1995), Watt (1972).

193.*Debregeasia hypoleuca* Wedd.

Syn : Boehmeria salicifolia Don.; *B. hypoleuca* Hochst.; *D. bicolor* Wedd. in DC.; *D. salicifolia* (D.Don) Rendle; *Missiessya hypoleuca* Wedd.; *Morocarpus salicifolius* Blume; *Urtica bicolor* Roxb.

Family : Urticaceae.

Hindi Names : **Hindi :** Sansarn, Siarn.

Distribution : W. Temperate Himalaya: 1,000-1,700m.

Part/s Used : Fruits.

Utilization : Ripe fruits used for flavouring food materials.

References : Bennett (1987), Kapur (1996a), Mehtani and Ingle (1999), Raju (2000), Watt (1972).

194.*Decalepis hamiltonii* Wight and Arn.

Family : Asclepiadaceae.

Distribution : Deccan Peninsula. Forest areas of W. Ghats.

Part/s Used : Roots. Bulbs.

Utilization : Both roots and bulbs used as condiments.

Active Principles : Aroma and taste of the root is due to the presence of 4-O-methylresorcylic aldehyde.

References : Ambasta (1986), Chopra *et al.* (1956), Mittre (1981).

195.*Decaspermum fruticosum* Forst.

Syn : D. paniculatum Kurz.

Family : Myrtaceae.

Distribution : Sikkim. Khasia Hills.

Part/s Used : Shoots.

Utilization : Terminal shoots employed as a condiment.

References : Ambasta (1986), Asolkar *et al.* (1992).

196.*Descurainia sophia* (L.) Webb. *ex* Prantl.

Syn : Sisymbrium sophia Linn.; *Sophia chirurgorum* Lobel.

Family : Brassicaceae.

English and Hindi Names: **Eng. :** Flixweed; **Hindi :** Khubkallana.

Distribution : Kashmir to Kumaon upto 4,500m. E. Himalaya.

Part/s Used : Seeds.

Utilization : Seeds added to food for flavouring.

Active Principles : Evobioside, evomonoside, erysimoside, helvetico-side and strophanthin isolated from seeds. Essential oil of seeds rich in benzyl and allyl isothiocyanates and allyl disulphide.

References : Bennett (1987), Kirtikar and Basu (1984), Negi *et al.* (1993), Rastogi and Mehrotra (1990, 1991, 1993).

197. *Desmos chinensis* Lour.

Syn : Unona discolor Vahl.

Family : Annonaceae.

Distribution : Forests of N.-E., S. and W. India.

Part/s Used : Flowers.

Utilization : Essential oil extracted from flowers used in perfumery.

References : Ambasta (1986), Bennett (1987), Chopra *et al.* (1956).

198. *Dianthus caryophyllus* Linn.

Family : Caryophyllaceae.

English Names: **Eng. :** Carnation, Clove Pink, Grenadine, Picotee.

Distribution : Cultivated all over India in gardens, especially on the hills.

Part/s Used : Flowers.

Utilization : Essential oil from flowers used in perfumery.

Active Principles : Pelargonidin-3-malylglucoside isolated.

References : Ambasta (1986), Rastogi and Mehrotra (1995), Watt (1972).

199. *Dillenia aurea* Smith

Syn : D. ornata Wall.; *D. pulcherrima* Kurz.; *D. speciosa* Griff.

Family : Dilleniaceae.

Hindi Name : **Hindi :** Aggai.

Distribution : Throughout plains of India.

Part/s Used : Fruits.

Utilization : Fruits used as seasoning materials and condiments.

Active Principles : Betulin, betulinic acid, lupeol and β-sitosterol isolated from fruits.

References : Ambasta (1986), Bennett (1987), Mittre (1981), Raju (2000), Rastogi and Mehrotra (1993), Watt (1972).

200. *Dillenia indica* Linn.

Syn : D. elliptica Thunb.; *D. speciosa* Thunb.

Family : Dilleniaceae.

English, Sanskrit and Hindi Names : **Eng. :** Dillenia; **Sans. :** Bhavya, Ruvya; **Hindi :** Aggai, Chalta, Girnar.

Distribution : Throughout Sub-Himalayan Tracts. Peninsular India, (Maharashtra, Mysore, Kerela, Andhra Pradesh, Orissa, Bihar and M.P.).

Part/s Used : Ripe fruits. Sepals.

Utilization : Ripe fruits employed to impart flavour to chutneys, jams and jellies. Thickened sepals have the flavour of apples, stewed and often mixed in prawn curries in Bengal.

Active Principles : Mucilaginous inner core of fruits contains arabinose (20.0), galactose (55.0), mannose (18.0%) and arabinogalactan (2%).

References : Asolkar *et al.* (1992), Kirtikar and Basu (1984), Mehtani and Ingle (1999), Rastogi and Mehrotra (1993), Sahni (1998), Watt (1972).

201. ***Dioscorea alata*** Linn.

Syn : D. atropurpurea Roxb.; *D. globosa* Roxb.; *D. purpurea* Roxb.; *D. rubella* Roxb.

Family : Dioscoreaceae.

English, Sanskrit and Hindi Names : **Eng. :** Asiatic Yam, Common Yam, Humped Yam, The Greater Yam, Wing Stalked Yam; **Sans. :** Pindalu; **Hindi :** Chupri alu.

Distribution : Cultivated in various parts of India.

Part/s Used : Tubers.

Utilization : Used for flavouring ice-creams.

Active Principles : Tubers yield alatanins A, B and C.

References : Bennett (1987), Kirtikar and Basu (1984), Mehtani and Ingle (1999), Rastogi and Mehrotra (1990, 1993, 1995), Watt (1972).

202. ***Diospyros tomentosa*** Roxb.

Syn : D. dubia Watt.; *D. melanoxylon* Roxb.; *D. wightiana* Wall.

Family : Ebenaceae.

English, Sanskrit and Hindi Names : **Eng. :** Coromandel Ebony, Malabar Ebony; **Sans. :** Dirghapatraka, Jalaja, Kakada, Kulaka; **Hindi :** Kendu, Temru, Tendu.

Distribution : Deciduous forests of Chota Nagpur, Bihar, W. Peninsula, Orissa.

Part/s Used : Flowers.

Utilization : Flowers fragrant.

References : Kirtikar and Basu (1984), Negi and Gaur (1994), Watt (1972).

203. ***Dipteryx odorata*** Will denow.

Family : Papilionaceae.

Part/s Used : Beans.

Utilization : Tonco beans find extensive use in perfumery, and frequently mixed with vanilla beans in the preparation of extract of vanilla.

Active Principles : Main fragrant components of seeds are coumarin (3%), and coumaric or ortho-oxycinnamic anhydride.

Reference : Greenish (1999).

204. ***Docynia indica*** Decne.

Syn.: Cydonia indica Spach.; *Pyrus inidca* Roxb.

Family: Rosaceae.

Distribution : E. Himalaya. Assam. Khasia Hills. Manipur.

Part/s Used : Fruits.

Utilization : Ripe fruit has a strong sweet smell like that of a quince.

References : Arora (1981), Watt (1972).

205. ***Dorema ammoniacum*** D.Don

Family : Apiaceae.

English Name : **Eng. :** The E. Giant Fennel.

Part/s Used : Roots. Resin.

Utilization : Roots used as incense and resin in perfumery.

Active Principles : Gum resin contains ammodoremin and chromandione – doremone A.

References : Ambasta (1986), Chevallier (1996), Greenish (1999), Rastogi and Mehrotra (1995), Watt (1972).

206. ***Draba setosa*** Royle

Family : Brassicaceae.

Distribution : N.W. Himalaya.

Part/s Used : Leaves.

Utilization : Powdered leaves mixed with ghee to make incense.

References : Bennett (1987), Singh *et al.* (1996).

207. ***Dracontomelum mangiferum*** Blume

Syn : D. dao (Blume) Merr. and Rolfe

Family : Anacardiaceae.

Distribution : Andaman Islands.

Part/s Used : Flowers. Leaves.

Utilization : Flowers and leaves used for flavourings.

Active Principles : (-) 1, 2, 3, 4, 6, 7 – hexahydro - 12H – indolo -(2,3-a) quinolizine isolated.

References : Ambasta (1986), Rastogi and Mehrotra (1990).

208.*Dryobalanops aromatica* Gaertn.f.

Syn : D. camphora Colebr.

Family : Dipterocarpaceae.

English and Hindi Names: **Eng. :** Barus camphor; **Hindi :** Barus Kapur, Bhimsaini-Kapur.

Part/s Used : Wood.

Utilization : Interior of the stem is a source of Borneo camphor. Essential oil from wood used to perfume soap.

Active Principles : Essential oil contains 30-35% of pinenes, camphene, diterpene, 9-10% (+)-borneol and terpineol and 20-25% sesquiterpenes.

References : Ambasta (1986), Asolkar *et al.* (1992), Harborne and Baxter (2001), Watt (1972).

209.*Durio zibethinus* Linn.

Family : Bombacaceae.

English Names : **Eng. :** Civet-Cat Fruit Tree, Durian.

Distribution : Cultivated in the Nilgiris on the W. Coast.

Part/s Used : Fruits.

Utilization : Flavour of the arils and ripe fruits exceedingly attractive and peculiar.

Active Principles : Fruits contain volatile flavouring constituent whereas aril contains about 23% sugars and 41mg/100g of vitamin C.

References : Ambasta (1986), Asolkar *et al.* (1992), Harborne and Baxter (2001), Sahni (1998), Watt (1972).

210.*Ehretia laevis* Roxb. var. ***aspera*** C.B. Clarke

Syn: E. aspera Willd.; *E. obtusifolia* Hochst *ex* DC.

Family : Ehretiaceae.

Distribution : Common throughout deciduous forests and in S. Andamans upto 900m.

Part/s Used : Flowers.

Utilization : Flowers faintly scented.

References : Asolkar *et al.* (1992), Kirtikar and Basu (1984), Negi and Gaur (1994), Singh and Pandey (1998), Watt (1972).

211.*Elaeagnus parvifolia* Wallich *ex* Royle

Syn: E. umbellata auct. non Thunb.: Hook.f.

Family : Elaeagnaceae.

Distribution : Temperate Himalaya 1,000-3,500m.

Part/s Used : Whole Plant.

Utilization : Plant strongly fragrant. Ripe fruits added to curries to impart taste and flavour.

References : Bennett (1987), Negi *et al.* (1993), Singh *et al.* (1996), Watt (1972).

212. ***Elaeagnus umbellata*** Thunb.

Family : Elaeagnaceae.

Distribution : Kashmir to Nepal upto 3,000m.

Part/s Used : Flowers.

Utilization : Flowers fragrant.

References : Negi and Gaur (1994), Chopra *et al.* (1980).

213. ***Elettaria cardamomum*** (L.) Maton (Pl. 7A)

Syn: Alpinia cardamomum Roxb.; *Amomum compactum* Roem. and Schult.

Family : Zingiberaceae.

English, Sanskrit and Hindi Names : **Eng.** : Malabar Cardamom, The Lesser Cardamom; **Sans.** : Bahula, Chandrika, Ela, Gaurangi, Upakunchika; **Hindi** : Choti-elachi, Ilayechi.

Distribution : Indigenous in W. and S. India. Wild in moist forest and hilly regions of Karnataka and Kerela.

Part/s Used : Seeds. Fruits.

Utilization : Cardamom (whole or ground or as seeds) used in perfumery and culinary preparations as a spice for flavouring cordials, bitters and other pharmaceutical preparations, artificial fruit flavours, spiced rice, meat, vegetable dishes, coffee, tea and to a small extent tobacco, baked goods including cakes, buns, pastries and bread, and also as an ingredient in curry powder, sausage products, soups, canned fish, etc. Seeds yield an essential oil, Oleum Cardamoni, find application in flavouring processed foods, beverages cordials, bitters and liquors.

Active Principles : Seed oil (average about 5%) yields α -pinene (1.9), sabinene (4.5), myrcene (0.2), limonene (14.3), cineol (930.7), cymene (1.9), Me heptenone (0.8), linalool (0.9), linalyl acetate (1.2), terpineol (0.8), α-terpineol (3.7), α -terpinyl acetate (28.1), borneol (0.1), neryl acetate (0.3), geraniol (0.7), nerol (1.4), neolidol (0.3), heptacosane (0.5) and unidentified compounds (9.7%).

References : Ambasta (1986), Anonymous (1999), Bhattacharjee (1998), Chevallier (1996), Greenish (1999), Jain (1968), Harborne and Baxter (2001), Kirtikar and Basu (1984), Kumar and Singh (2000), Mehtani and Ingle (1999), Rastogi and Mehrotra (1990, 1993, 1995), Watt (1972).

214.*Elsholtzia densa* Benth.

Family : Lamiaceae.

Distribution : Kashmir to Kumaon: 3,000-3,900m.

Part/s Used : Seeds.

Utilization : Seeds used as a spice.

References : Koelz (1979), Rastogi and Mehrotra (1995).

215.*Elsholtzia fruticosa* (D.Don) Rehder

Syn: E. polystachya Benth.; *Perilla fruticosa* D.Don.

Family : Lamiaceae.

Distribution : Himalaya from Kashmir to Sikkim and Khasia Hills: 1,200-2,700m.

Part/s Used : Flowering Tops.

Utilization : Flowering tops used for flavouring cordial drinks.

References : Bennett (1987), Mittre (1981), Rastogi and Mehrotra (1991, 1995).

216.*Eragrostiella bifaria* (Wight) Bor

Syn: Eragrostis bifaria Wight *ex* Steud.

Family : Poaceae.

Part/s Used : Roots.

Utilization : Roots used for flavouring Burmese cheroots.

Reference : Ambasta (1986).

217.*Eriodictyon californicum* Greene

Family : Hydrophyllaceae.

Part/s Used : Leaves.

Utilization : Oil used as a flavour to mask bitterness of quinine.

Active Principles : Extract of leaves contains eriodictyol, eriodicty-onone, tannins, resin and volatile oil.

Reference : Bhattacharjee (1998).

218.*Eruca sativa* Mill.

Syn: Brassica eruca Linn.; *B. erucoides* Roxb.; *E. vesicaria* (L.) Cav. ssp. *sativa* (Mill.) Thell.

Family : Brassicaceae.

English, Sanskrit and Hindi Names : **Eng.** : Rocket-Salad, Roquette, Tara-mira; **Sans.** : Bhutaghna, Daradharsha, Grahagna, Kadamba, Rajika, Sarishapa; **Hindi** : Seoha, Taramira.

Distribution : Cultivated as a cold weather crop in N.W. India and W. Himalaya upto 3,350m.

Part/s Used : Seeds.

Utilization : Seeds yield a pungent fixed oil with characteristic odour and used in pickles.

Active Principles : These are glucoerucin (seeds), linoleic (36.2) erucic (10.3%) acids and thiopentanonitrile (seed oil).

References : Ambasta (1986), Bennett (1987), Kirtikar and Basu (1984), Rastogi and Mehrotra (1991, 1995), Watt (1972).

219. ***Ervatamia divaricata*** (Linn.) Alston

Syn: E.coronaria Stapf.; *Tabernaemontana coronaria* R.Br.; *T. divaricata* (L.) R.Br. *ex* R. and S.

Family : Apocynaceae.

English, Sanskrit and Hindi Names : **Eng.** : East Indian Rosebay, Wax Flower; **Sans.** : Ashvathabheda, Gajapadapa, Kshiri, Nandi, Nandivriksha, Tagara, Vanaspati, Vishnupriya; **Hindi** : Chandni, Tagar, Taggai, Sugandabala.

Distribution : Upper Gengetic plain, Garhwal, E. Bengal, Khasia Hills, Assam, Hills of Vizagapatnam. Cultivated in many parts.

Part/s Used : Wood.

Utilization : Wood used in perfumery and as incense.

Active Principles : Indole alkaloids (I and II) isolated from the plant.

References : Ambasta (1986), Kirtikar and Basu (1984), Rastogi and Mehrotra (1991, 1995).

220. ***Ervatamia heyneana*** (Wall.) T. Cooke

Syn: Tabernaemontana heyneana Wall.

Family : Apocynaceae.

Distribution : Konkan. N. Kanara. W. Ghats in Malabar and Travancore up to 1,000 m.

Part/s Used : Wood.

Utilization : Wood used in perfumery and as incense.

Active Principles : Wood contains camptothecin (0.00013) and 9-methoxycamptothecin (0.00004%).

References : Ambasta (1986), Kirtikar and Basu (1984), Rastogi and Mehrotra (1991, 1993).

221. ***Eryngium foetidum*** Linn.

Family : Apiaceae.

Distribution : Assam.

Part/s Used : Whole Plant except inflorescence.

Utilization : Used as spice for flavouring rice and curries.

Active Principles : These are trans-2-dodecenal (45.5) and 2-dodecenoic acid (15.5%) (essential oil of aerial parts), 2,3,6-trimethylbenzaldehyde (37.55) and 2-formyl-1,1,5-trimethyl-cyclohexa-2,4-dien-6-ol (19.82) (root oil) and trans-2-dodecenal (59.72%) (leaf oil).

References : Ambasta (1986), Chandra (1997), Lalramnghinglova (2002), Rao *et al.* (1981), Saklani and Jain (1994), Rastogi and Mehrotra (1995).

222. ***Erythroxylum coca*** Lam.

Family : Erythroxylaceae.

English Names: **Eng. :** Bolivian Coca, Coca Bush, Cocaine Tree, Huanuco Coca, Spadic Bush.

Distribution : Native of warm countries.

Part/s Used : Flowers. Leaves.

Utilization : Fragrant flowers and decocainized extract of coca leaves used for flavouring ice-cream, candy, soft-drinks and liquors.

Active Principles : Coca leaves rich in cocaine, cinnamylescaine, β-truxilline tropacocaine, valerine, hygroline and cuscohygrine.

References : Bhattacharjee (1998), Kirtikar and Basu (1984), Rastogi and Mehrotra (1993), Watt (1972).

223. ***Erythroxylum monogynum*** Roxb.

Syn: E. indicum Bedd.; *Sethia indica* DC.

Family : Erythroxylaceae.

English Names: **Eng. :** Bastard Sandal, Deccany Deodar, Red Cedar.

Distribution : W. Ghats in dry hill forests of Travancore upto 1,000m.

Part/s Used : Wood.

Utilization : Wood used as incense. Wood oil akin to sandalwood, recommended for perfumery.

Active Principles : Wood yields (-)ent-8- (17), 13E-labdadien-15-ol and (-) hydroxydevadarool along with (+) allodevadarool, (+) devadarene, (+) devadarool, (+) monogynol, (+) hydroxymono-gynol, (+) hibaene epoxide, (+) hibaene, diterpenetriol A, erythroxylol B, (-) pimaridiene, (-) kauranol and kaurandiol.

References : Ambasta (1986), Kirtikar and Basu (1984), Mittre (1981), Rastogi and Mehrotra (1990, 1993), Watt (1972).

224. *Etlingera linguiformis* (Roxb.) R.M. Smith

Syn: Alpinia linguiforme Roxb.; *Amomum linguiforme* (Roxb.) Benth.

Family : Zingiberaceae.

Part/s Used : Rhizome.

Utilization : Rhizome oil finds application in perfumery.

Reference : Rehman and Yusuf (1996).

225. *Etlingera loroglossa* (Gagnep.) R.M. Smith

Syn: Achasma loroglossa (Gagnep.) K. Larsen; *Amomum loroglossum* Gagnep.

Family : Zingiberaceae.

Part/s Used : Rhizome.

Utilization : Oil from rhizomes used as perfume.

Reference : Rehman and Yusuf (1996).

226. *Eucalyptus citriodora* Hook.

Syn: E. maculata Hook. var. *citriodora* Bailey

Family : Myrtaceae.

English Name : **Eng. :** Lemon-Scented Eucalypt.

Distribution : Cultivated in many parts of India.

Part/s Used : Leaves.

Utilization : Essential oil extracted from leaves used in perfumery.

Active Principles : Leaves possess citronellal, citronellol and cineole in the ratio 72:27:1 alongwith α- and β- pinene, cineole, p-cymene, linalool, isopulegol, citronellyl acetate, geraniol, citriodorin, eucalyptin, β-sitosterol, betulinic acid and ursolic acid.

References : Ambasta (1986), Bhakuni (2000) Kirtikar and Basu (1984), Kumar and Singh (2000), Rastogi and Mehrotra (1991, 1993, 1995).

227. *Eucaluptus globulus* Labill. (Pl. 7B)

Family : Myrtaceae.

English Names: **Eng. :** Blue Gum Tree of Victoria and Tasmania.

Distribution : Nilgiris. Abbottabad. Ranikhet.

Part/s Used : Leaves. Twigs.

Utilization : Leaves and young shoots yield an essential oil, Blue gum eucalyptus oil, used as a fragrance component in soaps, detergents, lotions and perfumes, and also for flavouring baked goods, confectionery, meat products, alcoholic and non-alcoholic beverages.

Active Principles : Main constituents are gallic, caffeic, ferulic and protocatechiuc acids, quercitol, quercitin, rutin, hyperoside and quercitol-3-glucoside (leaves), 70-85% 1,8-cineole, α- and β- pinene, (+)-limonene, p-cymene, α-phellandrene, γ-terpinene, camphene aromadendrene and allo-aromadendrene (essential oil).

References : Bhakuni (2000), Chevallier (1996), Greenish (1999), Harborne and Baxter (2001), Kirtikar and Basu (1984), Rastogi and Mehrotra (1990, 1991, 1993, 1995), Watt (1972).

228. ***Eucalyptus radiata*** Sieb., *ex* DC.

Family : Myrtaceae.

English Name : **Eng. :** Narrow Leaved Peppermint.

Part/s Used : Leaves.

Utilization : Essential oil from leaves 3-3.5% used as flavour.

Active Principles : Oil constitutes cineole, geraniol, terpineol, citral and methyl cinnamate.

Reference : Bhattacharjee (1998).

229. ***Evernia prunastri*** (Linn.) Ach.

Family : Usneaceae.

Distribution : Jammu and Kashmir.

Part/s Used : Lichen Thallus.

Utilization : Much valued for scenting soaps and as a constituent of several perfumes.

Active Principles : Odorous constituents are α - and β- thujone, camphor, borneol and cineol.

References : Ambasta (1986), Mudgal and Hajra (1997).

230. ***Everniastrum cirrhatum*** (Fr.) Hale

Family : Parmeliaceae.

Part/s Used : Lichen Thallus.

Utilization : Used as spices to increase the taste and flavour of non-vegetarian preparations, curries and other vegetables.

References : Mudgal and Hajra (1997), Upreti *et al.* (2002).

231. ***Everniastrum nepalense*** (Taylor) Hale

Family : Parmeliaceae.

Part/s Used : Lichen Thallus.

Utilization : Used as spices.

Reference : Upreti *et al.* (2002).

232. ***Evodia lunu-ankenda*** Merrill

Syn: E. roxburghiana Benth.; *E. triphylla* Beddome; *Fagara triphylla* Roxb.; *Zanthoxylum triphyllum* Juss.

Family : Rutaceae.

Sanskrit Name : **Sans. :** Vanashampaga.

Distribution : W. Peninsula Khasia Hills, S. India, Andaman Islands.

Part/s Used : Leaves.

Utilization : Leaves used for flavouring.

References : Ambasta (1986), Kirtikar and Basu (1984), Watt (1972).

233. ***Farsetia jacquemontii*** Hook.f. and Thoms.

Syn: F. macrantha Blatter and Hallb.

Family : Brassicaceae.

Distribution : N.W. parts of India. Sandy places in the Punjab.

Part/s Used : Twigs.

Utilization : Tender twigs employed as a condiment.

References : Bennett (1987), Chandra (1997), Watt (1972).

234. ***Feijoa sellowiana*** Berg.

Syn: Acca sellowiana Berg.

Family : Myrtaceae.

English Names : **Eng. :** Feijoa, Guava, New Zealand Pine apple.

Part/s Used : Fruits.

Utilization : Fruits added to jams, jellies and beverages as a flavouring.

Active Principles : Fruits contain galenicals, ethyl butanoate, methyl and ethyl benzoates.

References : Asolkar *et al.* (1992), Mehtani and Ingle (1999).

235. ***Feronia limonia*** (Linn.) Swingle

Syn: Crataeva vallanga Koenig.; *F. elephantum* Correa.

Family : Rutaceae.

English, Sanskrit and Hindi Names : **Eng. :** Curd Fruit, Elephant Apple, Monkey Fruit, Wood Apple; **Sans. :** Akshasasya, Kapittha; **Hindi :** Bilin, Kait, Katbel, Kavitha, Kobitha.

Distribution : Sub-Himalayan forests. More plentiful in the moister tracts of Bombay, Madras and Bengal than in N. India.

Part/s Used : Leaves. Fruits.

Utilization : Crushed leaves added to food for flavouring. Aromatic fruits used to make chutney and sweet drinks.

Active Principles : Various constituents are psoralen, bergapten, orientin, vitexin and saponin (leaves), anisic acid, methyl chavicol

(27.2), trans-anethole (10.94), thymol (24.4) and p-cymen-7-ol (7.3%) (leaf essential oil), palmitic (15.2), oleic (43.4), linoleic (20.4) and linolenic acids (20.6%) (fruit oil).

References : Ambasta (1986), Anonymous (1999), Bajpayee and Dixit (1996), Chevallier (1996), Mittre (1981), Kirtikar and Basu (1984), Raju (2000), Rastogi and Mehrotra (1991, 1993, 1995), Upadhyay *et al.* (2000), Watt (1972).

236.*Ferula asafoetida* Linn. (Pl. 7C)

Syn: F. alliaceae Boiss.; *F. foetida* Regel.

Family : Apiaceae.

Sanskrit and Hindi Names: **Sans. :** Balhika, Hingu; **Hindi :** Hing.

Distribution : Kashmir.

Part/s Used : Latex or Gum-resin exuded from the living tap-root.

Utilization : Gum-resin employed to impart characteristic flavour in many food-stuffs, particularly curries, sauces, pickles, etc. Essential oil, i.e. Asafoetida oil used as a fragrance component in perfumes.

Active Principles : Essential oil contains 2-butyl propenyl disulphide, α - and β-pinene, ferulic acid, valeric acid and traces of vanillin. The components mainly responsible for the strong flavour are 2-butyl propenyl disulphide, 2-butyl-3-methyl-thioallyl disulphide, 1-(methylthiopropyl)-1-propenyl disulphide, 1-(1-methyl thio-1-propenyl disulphide, di-2-butyl trisulphide, 2-butyl methyl trisulphide, di-2-butyl disulphide, di-2-butyl tetrasulphide, 2-butyl methyl disulphide and dimethyl trisulphide.

References : Ambasta (1986), Bhattacharjee (1998), Chevallier (1996), Greenish (1999), Jain (1968), Harborne and Baxter (2001), Mehtani and Ingle (1999), Rastogi and Mehrotra (1991, 1993, 1995), Singh and Pandey (1998), Watt (1972).

237.*Ferula narthrex* Boiss.

Syn: Narthrex asafoetida Falc.

Family : Apiaceae.

English, Sanskrit and Hindi Names: **Eng. :** Asafoetida, Devil's Dung, Persian Fennel-Giant; **Sans. :** Agudagandha, Balhika, Hingu, Ramatha; **Hindi :** Hing, Hingda, Hingra.

Distribution : Native to Kashmir in India.

Part/s Used : Latex from tap-root.

Utilization : Another source of asafoetida used as a spice-like 'sambhar' in S. India to give a scented smell to the food preparations.

Active Principles : Organic sulphur isolated.

References : Ambasta (1986), Anonymous (1999), Asolkar *et al.* (1992), Bhattacharjee (1998), Chopra *et al.* (1980), Kirtikar and Basu (1984), Raju (2000), Watt (1972).

238.*Filipendula ulmaria* Maxim.

Family: Rosaceae.

English Names : **Eng. :** Medow sweet, Queen of the meadow.

Part/s Used : Flowers.

Utilization : Aroma of flowers resembles that of almonds.

Active Principles : Flowers yield glycosides of quercetin, phenolic glycosides (salicylates), volatile oil (salicylaldehyde) and polyphenols (tannins).

Reference : Chevallier (1996).

239.*Fissistigma rubiginosum* (A.DC.) Merrill.

Syn: Melodorum rubiginosum Hook.f. and Thoms.; *Uvaria rubiginosa* Hook.f. and Thoms.

Family : Annonaceae.

Part/s Used : Flowers.

Utilization : Flowers delightfully fragrant.

References : Ambasta (1986), Bennett (1987).

240.*Flacourtia jangomas* (Lour.) Raeusch.

Syn: F.cataphracta Roxb. *ex* Willd.; *Stigmarota jangomas* Lour.

Family : Flacourtiaceae.

English, Sanskrit and Hindi Names : **Eng. :** Many-Spined-Flacourtia, Puneala Plum; **Sans.:** Ahini, Kalameshi, Khadiraparni, Paniyamalaka, Talishapatra; **Hindi :** Paniala, Paniamalak, Paniaunvola, Talispatri.

Distribution : Commercially cultivated in India.

Part/s Used : Fruits.

Utilization : Fruit pulp used to give tarty flavour.

References : Ambasta (1986), Arora (1981), Bennett (1987), Kirtikar and Basu (1984), Negi and Gaur (1994), Raju (2000), Watt (1972).

241.*Flemingia vestita* Benth. *ex* Baker

Syn: F. procumbens Roxb.; *Moghania procumbens* (Roxb.) Mukerjee.

Family : Papilionaceae.

Distribution : Found in Himalaya; ascending to 2,900 m.

Part/s Used : Tubers.

Utilization : Outer skin of tubers has an agreeable flavour.

Active Principles : Yields genistein, daidzein, formononetin and pseudobaptigenin.

References : Bennett (1987), Negi and Pant (1990), Watt (1972).

242.*Foeniculum vulgare* Mill. (Pl. 7D)

Syn: Anethum foeniculum Linn.; *A. panmorium* Roxb.; *F. capillaceum* Gilib.; *F. dulce* C. Bauh.; *F. officinale* All.; *F. panmorium* DC.; *Ozodia foeniculacea* W. and A.

Family : Apiaceae.

English, Sanskrit and Hindi Names : **Eng.** : Fennel Fruit, Finkel Sweet Fennel; **Sans.** : Avakapushpi, Bhuripushpa, Madhurica, Madhavi; **Hindi** : Bari Saunf, Saunf, Sonp, Sont.

Distribution : Extensively cultivated in India upto 2,000m.

Part/s Used : Mericarps.

Utilization : Aromatic mericarps and fennel oil extracted from it are used to impart flavour and aroma to meat, soups, other dishes, sauces, bakery, confectionery products, cordials, liquors and pickles.

Active Principles : Anethol is the chief flavouring agent. Fennel oil (2.0-2.2%) contains α-pinene (3.0), camphene (0.65), α-phellan-drene (0.44), limonene (4.56), fenchone (10.20), methylchavicol (3.50), anethole (74.85), anisaldehyde (1.80) and p-anisic acid (1.0%).

References : Ambasta (1986), Anonymous (1999), Bhattacharjee (1998), Chevallier (1996), Chopra *et al.* (1960), Greenish (1999), Harborne and Baxter (2001), Kirtikar and Basu (1984), Mehtani and Ingle (1999), Rastogi and Mehrotra (1990, 1991, 1993, 1995), Watt (1972).

243.*Fragaria chiloensis* (L.) Duchesne

Syn: F. vesca auct. non Linn.

Family : Rosaceae.

English Name : **Eng.** : Garden strawberry.

Distribution : Found on the Khasia and Nilgiri mountains. Cultivated in Kumaon, H.P., Maharashtra.

Part/s Used : Fruits.

Utilization : Fruit esteemed as a dessert, and in ice-creams, soda-fountain beverages and strawberry wine.

Active Principles : Fruits contain isofucosterol (40.0%), sitosterol, campesterol, cholesterol and stigmasterol.

References : Ambasta (1986), Rastogi and Mehrotra (1990, 1995), Watt (1972).

244. *Garcinia cambogia* (Gaertn.) Desr.

Syn: G. affinis Wight and Arn.; *G. elliptica* Wall.; *G. gummi-gutta* (L.) Rob., Brittonia; *G. zeylanica* Roxb.

Family : Clusiaceae.

Distribution : Mountains of the W. Peninsula from Concan to Travancore.

Part/s Used : Rind.

Utilization : Dried rind of fruit eaten as a condiment in curries.

References : Ambasta (1986), Bennett (1987), Mittre (1981), Watt (1972).

245. *Garcinia cornea* Linn.

Syn: Diseostigma fabrile Miquel.; *G. affinis* Wall.

Family : Clusiaceae.

Distribution : E. Bengal.

Part/s Used : Fruits.

Utilization : Fruits with a pleasant flavour.

References : Ambasta (1986), Kirtikar and Basu (1984), Watt (1972).

246. *Garcinia hombroniana* Pierre

Family : Clusiaceae.

Distribution : Andaman and Nicobar Islands.

Part/s Used : Fruits.

Utilization : Fruits have flavour like that of peaches.

Active Principles : Bronianone isolated.

References : Ambasta (1986), Rastogi and Mehrotra (1991).

247. *Garcinia indica* (Dupetit-Thouars) Choisy

Syn: Brindonia indica Dupetit-Thouars.; *G.purpurea* Roxb.

Family : Clusiaceae.

English, Sanskrit and Hindi Names: **Eng. :** Brindonia Tallow Tree, Kokam Butter Tree, Mangosteen Oil Tree, Wild Mangosteen; **Sans. :** Amlabija, Bijamla, Chukramla, Puramla, Rasamla, Shakamla, Tintidika, Vrikshamla; **Hindi :** Kokam.

Distribution : Tropical rain forests of the W. Ghats, Malabar, the Konkan and Goa; often cultivated.

Part/s Used : Fruits. Seeds.

Utilization : Fruits used to impart flavour to curries, beverages, jellies and syrups. A valuable oil, 'Kokam Butter' obtained from seeds.

Active Principles : These are (-) hydroxycitric acid, cyaniding-3-glucoside, cyaniding 3-sambubioside (fruits), neutral lipids (88.0), and glycolipids (4.0%) (seeds).

References : Ambasta (1986), Kirtikar and Basu (1984), Mehtani and Ingle (1999), Rastogi and Mehrotra (1991, 1993), Sahni (1998), Watt (1972).

248. *Garcinia mangostana* Linn.

Family : Clusiaceae.

English and Hindi Names: **Eng.** : Mangosteen; **Hindi** : Mangusta, Mangustan.

Distribution : Cultivated in the W. Coast of Chennai, Nilgiris, Goa.

Part/s Used : Aril.

Utilization : Juicy white aril has flavour akin to the primrose.

Active Principles : Mangostin, three xanthones-gartanin, 8-deoxyg-artenin and normangostin isolated from fruits.

References : Ambasta (1986), Asolkar *et al.* (1992), Kirtikar and Basu (1984), Rastogi and Mehrotra (1991,1995), Sahni (1998), Watt (1972).

249. *Gardenia jasminoides* Ellis.

Syn: G. angusta (L.) Merrill; *G. florida* Linn.; *Varneria angusta* L., *Amoen.* Acad.

Family : Rubiaceae.

English, Sanskrit and Hindi Names : **Eng.** : Cape Jasmine; **Sans.** : Gandharaja; **Hindi** : Gandharaj.

Distribution : Often cultivated in Indian gardens.

Part/s Used : Flowers.

Utilization : Flowers contain an essential oil valued for perfuming tea.

Active Principles : Flowers possess two cycloartane triterpenoids – gardenic acid A and gardenic acid B.

References : Ambasta (1986), Bennett (1987), Kirtikar and Basu (1984), Rastogi and Mehrotra (1995), Watt (1972).

250. *Gardenia taitensis* DC.

Family : Rubiaceae.

Part/s Used : Flowers.

Utilization : Fragrant flowers used for perfume making.

Reference : Vickery (1990).

251. *Gastrochilus pandurata* Ridley

Family : Zingiberaceae.

Distribution : Andamans. Konkan, Archipelago, Java.

Part/s Used : Bulbs.

Utilization : Bulbs used as condiments.

References : Mittre (1981), Kirtikar and Basu (1984).

252. *Gaultheria fragrantissima* Wall.

Family : Ericaceae.

English, Sanskrit and Hindi Names: **Eng. :** Fragrant Wintergreen, Indian Wintergreen; **Sans. :** Hemant-harit; **Hindi :** Gandhpura-ka-tel.

Distribution : C. and E. Himalaya. Khasia Hills. Hills of S. India.

Part/s Used : Leaves.

Utilization : Leaves source of Indian Wintergreen oil, employed for flavouring soft drinks, lemon drops, tooth pastes, confectionery, etc.

Active Principles : Essential oil from fresh leaves (0.7-0.8%) yields methyl salicylates (99.6%), β-sitosterol, quercetin-3-galactoside and ursolic acid.

References : Ambasta (1986), Asolkar *et al.* (1992), Jain (1968), Kirtikar and Basu (1984), Mehtani and Ingle (1999), Rastogi and Mehrotra (1991).

253. *Gaultheria procumbens* Linn.

Family: Ericaceae.

English Names : **Eng. :** Canada Tea, Checkberry, Creeping Wintergreen.

Distribution : Introduced into India.

Part/s Used : Leaves.

Utilization : Aromatic leaves source of Wintergreen oil, used as a flavouring agent and in perfumes.

Active Principles : Oil constitutes 99% methyl salicylate.

References : Asolkar *et al.* (1992), Bhattacharjee (1998).

254. *Gelsemium sempervirens* (L.) Ait. f.

Family : Loganiaceae.

English Names: **Eng. :** Caroline Yellow Jasmine, False Jasmine.

Part/s Used : Flowers.

Utilization : Flowers fragrant.

Reference : Bhattacharjee (1998).

255. *Geum urbanum* Linn.

Family : Rosaceae.

English Names: **Eng. :** Avens, Blessed Herb, Herb Bennet, Way Bennet, Wild Rye, Wood Avens.

Distribution : Temperate W. Himalaya (2,000-3,500m) - Kashmir to Kumaon.

Part/s Used : Leaves.

Utilization : Leaves used for flavouring ale.

Active Principles : Main constituents of leaves are phenolic glycosides (including eugenol), tannins, a volatile oil, and possibly a sesquiterpene lactone (cnicin).

References : Ambasta (1986), Chevallier (1996), Kirtikar and Basu (1984), Watt (1972).

256. ***Globba marantina*** Linn.

Syn: G. bulbifera Roxb.

Family : Zingiberaceae.

Distribution : Khasia Hills. Meghalaya.

Part/s Used : Bulbils. Flowers.

Utilization : Bulbils spicy and used for seasoning by hilly tribals. Flowers fragrant.

References : Ambasta (1986), Asolkar *et al.* (1992), Bennett (1987), Chandra (1997), Rehman and Yusuf (1996).

257. ***Glycosmis cochinchinensis*** Pierre *ex* Engler

Syn: G. arborea (Roxb.) DC.; *G. chylocarpa* W. and A.; *G. pentaphylla* Correa; *G. retzii* Roem.; *Limonia arborea* Roxb.; *L. pentaphylla* Rets.; *Myxospermum chylocarpum* Roem.

Family : Rutaceae.

Sanskrit and Hindi Names: **Sans.** : Ashvashakota; **Hindi** : Bannimbu, Girgitti, Pilrupotala, Potali.

Distribution : Throughout the Tropical and Sub-Tropical Himalaya, ascending to 2,350m in Sikkim.

Part/s Used : Flowers. Fruits.

Utilization : Flowers fragrant. Ripe fruits flavoury in taste.

Active Principles : Flower-heads contain glycophymine and amide glycomide.

References : Asolkar *et al.* (1992), Bennett (1987), Kirtikar and Basu (1984), Negi and Gaur (1994), Watt (1972).

258. ***Glycyrrhiza glabra*** Linn. (Pl. 7E)

Family : Papilionaceae.

English, Sanskrit and Hindi Names : **Eng.** : Licorice, Liquorice; **Sans.** : Jalayashti, Madhuka, Yashti-Madhu; **Hindi:** Jethi-Madh, Mulhatti.

Distribution : Met with in Kashmir Himalaya. Cultivated especially in Sub-Tropical and Tropical regions.

Part/s Used : Rhizomes. Roots.

Utilization : Rhizomes and roots and liquorice extract containing ammoniated glycyrrhizin, used mainly as a sweetening agent, giving aroma to beer and for flavouring cigarettes, snuff, tobacco, confectionery, beverages, jams and marmalades, sweet sauces, reconstituted vegetable proteins, chocolates, chewing gum and diet colas. Also, yield an essential oil.

Active Principles : Characteristic aroma of the volatile oil of the liquorice is due to estragole, anethole, eugenol, indole, γ-nonalactone and cumin alcohol. Roots contain triterpenoids, glycyrrhisoflavanone, glycyrrhisoflavone, licuroside, glycyrrhizin, glycyrrhizic acid whereas hexanoic acid (31.57), γ-nonalactone (1.33), cumic alcohol (0.79), indole (0.25), anethole (0.23), eug-enol (0.21) and estragole (0.18%) isolated from root essential oil.

References : Ambasta (1986), Anonymous (1999), Bhattacharjee (1998), Chevallier (1996), Chopra *et al.* (1960), Greenish (1999), Kirtikar and Basu (1984), Mehtani and Ingle (1999), Rastogi and Mehrotra (1991, 1993, 1995), Sharma and Sood (1997), Sinha and Sinha (2001), Watt (1972).

259. *Gonystylus macrophyllus* (Miq.) Airy Shaw

Syn: G. bancanus Baill.; *G. miquelianus* Teijsm and Binn.

Family : Thymelaeaceae.

Distribution : Nicobar Islands.

Part/s Used : Wood.

Utilization : Scented wood and its volatile oil used in incenses.

Active Principles : Wood oil contains gonystylol.

References : Ambasta (1986), Chopra *et al.* (1980).

260. *Guettarda speciosa* Linn.

Syn: Cabamba jasminiflora Sonner.; *Jasminum hirsutum* Willd.; *Nictanthes hirsuta* Linn.

Family : Rubiaceae.

Distribution : Tidal forests along the shores of S. India. Andaman Islands.

Part/s Used : Flowers.

Utilization : Extract of fragrant flowers akin to rose water, much valued in Kerela.

References : Ambasta (1986), Chopra *et al.* (1980), Watt (1972).

261. *Gynandropsis gynandra* (Linn.) Briq.

Syn: Cleome gynandra L.; *C. pentaphylla* Linn.; *G. pentaphylla* (L.) DC.

Family : Capparidaceae.

English, Sanskrit and Hindi Names: **Eng. :** Dog Mustard; **Sans. :** Ajagandha, Arkapushpika, Brahmi, Choraka, Kabari, Kharapushpa, Surjavarta, Ugragandha; **Hindi :** Churota, Gandhuli, Hulhul, Hurhur, Karalai, Lalhulhul, Safedhulhul.

Distribution : Throughout the warmer parts of India.

Part/s Used : Leaves.

Utilization : Leaves used for flavouring sauces and curries.

Active Principles : Leaves contain centaureidin, kaempferol and its 3-O-diglucoside, quercitrin, myricitrin, α- and β- amyrins, taraxasterol, β-sitosterol and capric, lauric, myristic, palmitic, palmitoleic, stearic, oleic and linoleic acids.

References : Ambasta (1986), Anonymous (1999), Kirtikar and Basu (1984), Rastogi and Mehrotra (1991,1995), Watt (1972).

262.*Hedychium acuminatum* Rosc.

Syn: H. spicatum var. *acuminatum* (Roscoe) Wall.

Family : Zingiberaceae.

Sanskrit and Hindi Names: **Sans. :** Gandhashati, Karpurakachali; **Hindi :** Kapura-Kachari.

Distribution : Himalaya: Garhwal to Sikkim upto 2,400m.

Part/s Used : Roots.

Utilization : An incense (dhoop) prepared from its root.

Active Principles : Ethanolic extract of plant contains β-sitosterol and its glucoside.

References : Asolkar *et al.* (1992), Bennett (1987), Negi *et al.* (1999), Rastogi and Mehrotra (1991).

263.*Hedychium coronarium* Koenig. var. ***flavum*** Baker

Syn: H. flavum Roxb.

Family : Zingiberaceae.

English Names: **Eng. :** Common Ginger Lily, Garland Flower, Yellow Ginger Lily.

Distribution : Throughout moist parts of India upto 2,400m. Also grown in gardens.

Part/s Used : Flowers.

Utilization : Fragrant flowers used in high grade perfumes.

Active Principles : Linalool, methyl benzoate, cis-jasmone, eugenol, (E) isoeugenol, jasmin lactone, methyl jasmonate and methyl epijasmonate isolated from flower oil.

References : Ambasta (1986), Asolkar *et al.* (1992), Rastogi and Mehrotra (1990, 1995), Rehman and Yusuf (1996).

264.*Hedychium spicatum* Buch.-Ham.

Family : Zingiberaceae.

English, Sanskrit and Hindi Names : **Eng. :** Spiked Ginger Lily, The Lesser Galangal of Ainslie; **Sans. :** Amlaharidra, Durva, Gandha, Karpurakachali, Himaja, Palashi; **Hindi:** Gandhapalashi, Kapurakachari, Sitruti.

Distribution : Sub-Tropical Himalaya.

Part/s Used : Rhizome. Root.

Utilization : Aromatic rhizomes employed in the preparation of 'Abir'- a fragrant, coloured powder, used during the Holi festival and in religious ceremonies. Also, its oil used in perfumery for preparation of 'Attar'. Roots used in perfume, soap, hair oil, face powder, etc.

Active Principles : Rhizomes contain hedychenone, diterpene-6-oxo-labda-7,11,13-trien-16-oic acid lactone (I).

References : Ambasta (1986), Anonymous (1999), Kirtikar and Basu (1984), Misra and Kumar (2000), Purohit (2000), Rastogi and Mehrotra (1993, 1995), Sarin (1989), Singh (1999), Watt (1972).

265. *Heliotropium arborescens* Linn.

Syn: H. peruvianum Linn.

Family : Boraginaceae.

English Names : **Eng. :** Cherry Pie, Common Heliotrope.

Distribution : Grown as ornamental.

Part/s Used : Flowers.

Utilization : Flowers yield a perfume.

References : Ambasta (1986), Asolkar *et al.* (1992).

266. *Hemerocallis fulva* Linn.

Family : Liliaceae.

English Names: **Eng. :** Orange Day-Lily, Tawny Day-Lily.

Distribution : Himalaya and Khasia Hills, grown throughout India in gardens.

Part/s Used : Flowers.

Utilization : Flowers used to flavour foods.

Active Principles : Aqueous extract of plant yields choline.

References : Ambasta (1986), Rastogi and Mehrotra (1995).

267. *Hemidesmus indicus* (L.) R.Br.

Syn: Asclepias pseudosarsa Roxb.; *H.wallichii* Miq.; *Periploca indica* Willd.

Family : Asclepiadaceae.

English, Sanskrit and Hindi Names : **Eng. :** False Sarsaparilla, Indian Sarsaparilla; **Sans. :** Anantamula, Bhadra, Bhadravalli, Sariva; **Hindi :** Anantamul, Hindisalsa, Janglichanbelli, Magrabu.

Distribution : Upper Gangetic Plains. S. India.

Part/s Used : Roots.

Utilization : The roots used for flavouring sweet smelling drink in place of coffee and tea, sherbet and syrup.

Active Principles : Major constituents are p-methoxy salicylaldehyde (plant), hexatriacontane, lupeol, its octacosanoate, α-amyrin, β-amyrin, its acetate, β-sitosterol, two coumarinolignoids-hemidesmin 1 and hemidesmin 2 (roots).

References : Anonymous (1999), Chopra *et al.* (1980), Greenish (1999), Kirtikar and Basu (1984) Mehtani and Ingle (1999), Prasad *et al.* (1996), Rastogi and Mehrotra (1991), Sinha and Sinha (2001), Watt (1972).

268.*Heracleum lanatum* Michx.

Syn: H. candicans Wall. *ex* DC.; *H. nepalense* D.Don.

Family : Apiaceae.

Distribution : Kashmir to Kumaon: 700-1,350m.

Part/s Used : Fruits.

Utilization : Fruits used as spices.

Active Principles : Two lactones and a furanocoumarin – bergapten isolated from fruits.

References : Chopra *et al.* (1980), Hajra and Chakraborty (1981), Rastogi and Mehrotra (1990,1991).

269.*Hesperethusa crenulata* (Roxb.) M. Roem.

Syn: Limonia acidissima auct., non Linn.; *L. crenulata* Roxb.; *Naringi crenulata* (Roxb.) Nicolson.

Family : Rutaceae.

Hindi Name : **Hindi :** Beli.

Distribution : Throughout India.

Part/s Used : Fruits.

Utilization : Fruits occasionally used as a condiment for flavouring fish, meat and meat products, etc.

Active Principles : Fruits yield acidissimin, acidissiminol, acidissi-minin and 2,6-dimethoxybenzoquinone.

References : Ambasta (1986), Chopra *et al.* (1980), Rastogi and Mehrotra (1991, 1995).

270.*Heterodermia tremulans* (Mull.Arg.) Culb.

Family : Physciaceae.

Part/s Used : Lichen Thallus.

Utilization: Used as spices and for flavouring non-vegetarian preparations, pulses and other vegetables.

References : Mudgal and Hajra (1997), Upreti (1996).

271. *Hibiscus sabdariffa* Linn. var. ***sabdariffa***

Family : Malvaceae.

English and Hindi Names: **Eng. :** Indian Sorrel, Jamaica Sorrel, Natal Sorrel, Red Sorrel, Roselle Hemp Plant; **Hindi :** Lal-ambari, Patwa.

Distribution : Cultivated in the hotter parts of India.

Part/s Used : Calyx. Leaf-stalks.

Utilization : Tender leaves and stalks used for seasoning curries. Fleshy calyx used for flavouring jellies, sauces, syrups, wine and confectionery.

Active Principles : Sepals contain delphinidin-3-O-sambubioside, cyaniding-3-O-glucopyranoside, quercetin, myricetin, hibiscetin, hibiscitrin, O-coumaric and p-coumaric acids. Leaves yield sitosterol and β-D-galactoside.

References : Ambasta (1986), Asolkar *et al.* (1992), Kirtikar and Basu (1984), Mehtani and Ingle (1999), Rastogi and Mehrotra (1995), Watt (1972).

272. *Hippophae rhamnoides* Linn.

Syn: H. tibetana Schlecht.

Family : Elaeagnaceae.

English Names : **Eng. :** Common Seabuck Thorn, Sallow-Thorn.

Distribution : W. Himalaya.

Part/s Used : Fruits.

Utilization : Ripe fruits juicy with an agreeable flavour and used to make wine, beer, jams, condensed juice, cheese, candies, butter, soda water.

Active Principles : Fruits yield isorhamnetin-3-β-D-glucoside, isorhamnetin-3-rutinoside, isorhamnetin-3-β-D-glucosido α–L-rhamnoside, isorhamnetin-3-β-sophorosido-7-L-rhamnoside, isorhamnetin tetroside, rutin, quercetin, quercetin-7-O-rhamnoside, quercetin-3-O-methyl ether, 3,4-dihydroxybenzoic and p-coumaric acids.

References : Ambasta (1986), Chevallier (1996), Duhoon *et al.* (1996), Kirtikar and Basu (1984), Rastogi and Mehrotra (1990, 1991, 1993), Sharma and Sood (1997), Singh *et al.* (1996), Watt (1972).

273. *Holboellia latifolia* Wall. var. ***latifolia*** Hook.f.

Syn: H. acuminata Lindl.; *Stauntonia latifolia* Wall.

Family : Lardizabalaceae.

Distribution : Himalaya (1,350-3,000m): Kumaon eastward to Bhutan. Khasia Hills. Upper Assam at low elevations.

Part/s Used : Climbers.

Utilization : Climbers scented.

References : Negi and Gaur (1994), Watt (1972).

274. *Homalomena rubescens* Kunth.

Syn: Calla rubescens Roxb.

Family : Araceae.

Distribution : Sikkim Himalaya. Khasia Hills.

Part/s Used : Rhizome.

Utilization : Oil from the rhizomes used in perfumery.

Active Principles : Rhizome oil possesses linalool (60%), α -terpineol and linalyl acetate.

References : Ambasta (1986), Chopra *et al.* (1980), Kirtikar and Basu (1984), Rastogi and Mehrotra (1995).

275. *Horsfieldia irya* (Gaertn.)Warb.

Syn: Myristica exaltata Wall.; *M. irya* Gaertn.; *M. javanica* Blume.; *M. sphaerocarpa* Wall.

Family : Myristicaceae.

Distribution : Andaman Islands.

Part/s Used : Flowers.

Utilization : Fragrant flowers yield oil used in perfumery.

References : Ambasta (1986), Bennett (1987), Watt (1972).

276. *Houttuynia cordata* Thunb.

Family : Piperaceae.

English Name : **Eng. :** Tsi.

Distribution : Himalaya, H.P. to Sikkim.

Part/s Used : Leaves. Roots.

Utilization : Dried leaves and roots used as spices.

Active Principles : Active ingredients are geraniol, borneol, humulene, thymol, octanal, decanal, dodecanal, decanoic, dodecanoic, tetradecanoic, hexadecanoic and afzelin, hyperin, rutin, β-sitosterol, chlorogenic, palmitic, linoleic, oleic and stearic acids (aerial parts), quercitrin and inorganic salts like potassium chloride and potassium sulphate (root extract).

References : Asolkar *et al.* (1992), Chopra *et al.* (1980), Negi *et al.* (1993), Negi and Pant (1990), Rastogi and Mehrotra (1991, 1993).

277. ***Humulus lupulus*** Linn.

Family : Moraceae.

English Name : **Eng.** : The Common Hop.

Distribution : Cultivated in Lahaul region.

Part/s Used : Inflorescence.

Utilization : Extracted Hop oil from glandular hairs of the female inflorescense (Hops) used for giving aroma and flavour to beer, cereal beverages, mineral water, tobacco and in some perfume compositions.

Active Principles : Hop oil rich in β-pinene, myrcene, α-humulene and β-humulene, 2,3,4-trithiapentane.

References : Ambasta (1986), Chevallier (1996), Greenish (1999), Harborne and Baxter (2001), Kirtikar and Basu (1984), Rastogi and Mehrotra (1990, 1991, 1993, 1995), Sharma and Sood (1997), Singh *et al.* (1996), Watt (1972).

278. ***Hydrodictyon reticulatum*** (Linn.) Lagerh.

Family : Hydrodictyaceae.

Part/s Used : Algal Thallus.

Utilization : Algal thallus cooked with vegetables for its charact-eristic fishy smell.

Reference : Ambasta (1986).

279. ***Hyphaene thebaica*** Mart.

Family : Arecaceae.

English Name : **Eng.** : Egyptian Doum Palm.

Distribution : Grown in Indian gardens.

Part/s Used : Fruit.

Utilization : Fruit stones used in perfumery.

References : Ambasta (1986), Rastogi and Mehrotra (1991).

280. ***Hyptis pectinata*** (L.) Poit.

Family : Lamiaceae.

Distribution : Assam, W. Bengal, Tamil Nadu.

Part/s Used : Leaves. Resin.

Utilization : Leaves used for flavouring soups and resin used as incense.

Active Principles : Leaf oil contains p-cymene, thymol, γ-terpinene, myrcene and α -thujene.

References : Ambasta (1986), Mehtani and Ingle (1999), Rastogi and Mehrotra (1995).

281. *Hyptis suaveolens* (L.) Poit.

Family : Lamiaceae.

Hindi Name : **Hindi :** Wilayati Tulsi.

Distribution : Deccan Peninsula. N.E. India. Andaman Islands.

Part/s Used : Shoot Tops.

Utilization : Shoot tops employed for flavouring foods. Yield of volatile oil is 0.06%.

Active Principles : Essential oil of plant yields 1,8-cineole (33.4%), others are α-pinene, myrcene and β-caryophyllene.

References : Kirtikar and Basu (1984), Mehtani and Ingle (1999), Rastogi and Mehrotra (1993, 1995).

282. *Hyssopus officinalis* Linn.

Family : Lamiaceae.

English Name : **Eng. :** Hyssop.

Distribution : W. Himalaya from Kashmir to Kumaon: 2,600-3,700m.

Part/s Used : Leaves. Flower Tops.

Utilization : Leaves and flowering tops used as spice and for flavouring salads, soups, liquors. Hyssop oil extracted from herb also used to flavour pickles, meat, sauces, alcoholic beverages

Active Principles : Pinocamphone (69.1), β-pinene (16.8), camphene (11.5) and pinocampheol (1.1%) isolated from essential oil (1.0-1.5%) of aerial parts.

References : Ambasta (1986), Bhattacharjee (1998), Chevallier (1996), Harborne and Baxter (2001), Kirtikar and Basu (1984), Mehtani and Ingle (1999), Negi *et al.* (1993), Rastogi and Mehrotra (1991, 1993, 1995), Sharma and Sood (1997), Sinha and Sinha (2001), Watt (1972).

283. *Ilex vomitoria* Ait.

Syn: I. cassine Walt. non Linn.

Family : Aquifoliaceae.

English Names : **Eng. :** Cassena, Yaupon.

Distribution : Cultivated in Indian gardens.

Part/s Used : Leaves.

Utilization : Leaves used for preparing a tea-like beverage (black drink), and syrup for use in soft-drinks.

Active Principles : Caffeine.

References : Ambasta (1986), Asolkar *et al.* (1992), Mehtani and Ingle (1999).

284. *Illicium griffithii* Hook. f. and Thoms.

Family : Magnoliaceae.

Distribution : Khasia Hills: 1,350-2,700m.

Part/s Used : Fruits.

Utilization : Aromatic fruits with flavour between that of cubebs and bay leaves used as a spice sauce and pickle foods.

References : Bajpayee and Dixit (1996), Dam and Hajra (1981), Kirtikar and Basu (1984), Saklani and Jain (1994), Watt (1972).

285. *Inula racemosa* Hook.f.

Syn: I. helenium Herb.

Family : Asteraceae.

Distribution : Temperate and Alpine Himalaya: 1,700-4,750m.

Part/s Used : Roots.

Utilization : Essential oil from aromatic roots used as a flavour ingredient in alcoholic and non-alcoholic beverages, confectionery, baked goods, gelatins and puddings. Also, used as a fragrance component in cosmetic products and oriental type perfumes.

Active Principles : Roots contain 10% inulin, alantolactone, isoalantolactone, dihydroisoalantolactone and its oil rich in alantol-actone, inunolide, allantolides – I, II, III, isoinunal and telekin.

References : Ambasta (1986), Harborne and Baxter (2001), Kaul (1997), Kirtikar and Basu (1984), Rastogi and Mehrotra (1990, 1993, 1995), Sood *et al.* (2001), Watt (1972).

286. *Iris germanica* Linn.

Syn: I.deflexa Knowles and Westc.; *I. nepalensis* Wall.

Family : Iridaceae.

English, Sanskrit and Hindi Names : **Eng. :** Iridis Rhizoma, Orris Root; **Sans. :** Padma-pushkara; **Hindi :** Keore-ka-mul.

Distribution : Cultivated in Kashmir.

Part/s Used : Rhizome.

Utilization : Orris rhizome used for flavouring and as a perfuming agent. Yields an essential oil used in perfumery.

Active Principles : Irone, an oily liquid with a powerful odour of violets is the only aromatic constituent of the rhizome.

References : Ambasta (1986), Chopra *et al.* (1956, 1980), Greenish (1999), Watt (1972).

287. *Jasminum angustifolium* Vahl

Syn: J. vimineum Willd.; *J. triflorum* Pers.; *Mogorium triflorum* Lamk..; *M. vimineum* Lamk.; *Nyctanthes angustifolia* Linn.; *N. triflora* Burm.; *N. viminea* Rets.

Family : Oleaceae.

English, Sanskrit and Hindi Names : **Eng.** : Wild Jasmine; **Sans.** : Asphota, Kananamallica, Priya, Supuja, Vanamalli; **Hindi** : Banmallika, Mwari.

Distribution : Lower hills of Deccan.

Part/s Used : Flowers.

Utilization : Flowers emit delightful perfume.

Active Principles : Flower contains indole.

References : Ambasta (1986), Asolkar *et al.* (1992), Kirtikar and Basu (1984), Watt (1972).

288. *Jasminum auriculatum* Vahl.

Family : Oleaceae.

Sanskrit and Hindi Names : **Sans.** : Ambashtha, Ganika, Juthika, Mugdhee, Suchimallika; **Hindi** : Juhi, Jui.

Distribution : Deccan, W. Peninsula.

Part/s Used : Flowers.

Utilization : Flowers employed for the production of perfumed hair oils and attars.

Active Principles : Cis-jasmone and dihydrojasmone are the main constituents.

References : Ambasta (1986), Kirtikar and Basu (1984), Rastogi and Mehrotra (1993, 1995).

289. *Jasminum flexile* Vahl.

Syn: J. azoricum L.

Family : Oleaceae.

Distribution : Deccan Peninsula.

Part/s Used : Flowers.

Utilization : Cultivated for its fragrant flowers.

Active Principles : A seco-iridoid-9'-deoxyjasminigenin isolated from flowers.

References : Ambasta (1986), Chopra *et al.* (1956), Rastogi and Mehrotra (1995).

290. *Jasminum humile* Linn.

Syn: J. bignoniaceum Wall. *ex* G.Don.; *J. chrysanthemum* Roxb.; *J. inodorum* Jacq.; *J. pubigerum* D.Don var. *glabram* DC.; *J. revolutum* Sims.; *J. wallichianum* Lindl.

Family : Oleaceae.

English, Sanskrit and Hindi Names : **Eng. :** Golden Jasmine, Italian Jasmine, Nepal Jasmine, Yellow Jasmine; **Sans. :** Gandadhya, Haima, Hemapushpika, Manohara, Svarnajuthika; **Hindi :** Malto, Peelajui, Peeli chameli, Pitmalti.

Distribution : Widely cultivated in gardens throughout India.

Part/s Used : Flowers.

Utilization : Flowers used in high grade perfumes, expensive soaps, cosmetics, mouth washes, dentifrices, bath salts, sachets, perfumed tobacco and also in incenses.

Active Principles : Flavanoids.

References : Ambasta (1986), Asolkar *et al.* (1992), Bennett (1987), Kirtikar and Basu (1984), Raju (2000), Watt (1972).

291. *Jasminum lanceolarium* Roxb.

Syn: J. paniculatum Roxb.

Family : Oleaceae.

Part/s Used : Flowers.

Utilization : Flowers used for scenting tea.

Reference : Ambasta (1986).

292. *Jasminum multiflorum* (Burm.f.) Andrews.

Syn: J.pubescens (Retz.) Willd.; *Nyctanthes multiflora* Burm.f.

Family : Oleaceae.

Sanskrit and Hindi Names: **Sans. :** Kunda, Vasanta; **Hindi :** Chameli, Kundphul.

Distribution : Throughout India.

Part/s Used : Flowers.

Utilization : Flowers sweet scented and used in perfumery.

Active Principles : Aerial parts yield seco-iridoid lactones-jasmo-lactones A, B, C and D and secoiridoid glucoside-jasmultiside.

References : Bennett (1987), Chopra *et al.* (1956), Raju (2000).

293. *Jasminum odoratissimum* Linn.

Family : Oleaceae.

English Name : **Eng. :** Jasmine.

Distribution : Introduced into Indian gardens.

Part/s Used : Flowers.

Utilization : Flowers used for flavouring tea.

Active Principles : Linalool, methyl benzoate, benzyl acetate, benzyl alcohol, cis-jasmone, eugenol, cis-3-hexenyl benzoate, vanillin, nerolidol, geranyl linalool, iso-phytol and phytol isolated from flowers.

References : Ambasta (1986), Rastogi and Mehrotra (1991,1993).

294.*Jasminum officinale* Linn. forma ***grandiflorum*** (Linn.) Kobuski (Pl. 8A)

Syn: J. aureum Don; *J. grandiflorum* Linn.; *J. officinale* L. var. *grandiflorum* Bailey

Family : Oleaceae.

English, Sanskrit and Hindi Names: **Eng. :** Common Jasmine, Spanish Jasmine, White Garden Jasmine, White Jasmine; **Sans.:** Ambashtha, Bahugandha, Chambeli, Chetaki, Jati, Malati, Vasanti; **Hindi :** Chamba, Chameli, Jati.

Distribution : Himalaya 1,000-3,000m. Often cultivated in India.

Part/s Used : Flowers.

Utilization : Jasmine obtained from flowers used in fragrance formulations in cosmetics, floral and oriental fragrances; and as a flavouring ingredient in major food products, alcoholic and non-alcoholic beverages, confectionery, baked goods, gelatins, puddings and frozen dairy desserts.

Active Principles : Oil contains benzyl acetate, indole, cis-jasmone, methyl jasmonate, linalool, phenylacetic acid, benzyl alcohol, farnesol, methyl anthranilate, jasmolactone, methyldehydro-jasmonate, cis- and trans- jasmonates.

References : Ambasta (1986), Asolkar *et al.* (1992), Bennett (1987), Bhakuni (2000), Harborne and Baxter (2001), Kirtikar and Basu (1984), Kumar and Singh (2000), Raju (2000), Rastogi and Mehrotra (1990, 1991, 1993), Watt (1972).

295.*Jasminum sambac* (Linn.) Ait.

Syn: J. fragrans Salisb.; *J. pubescens* Wall. not of Willd.; *J. quinque-florum* Heyne.; *J. undulatum* Willd.; *J. zambac.* Roxb.; *Mogorium sambac* Lamk.; *M. undulatum* Lamk.; *Nyctanthes sambac* Linn.; *N. undulatum* Linn.

Family : Oleaceae.

English, Sanskrit and Hindi Names: **Eng. :** Arabian Jasmine, Lily Jasmine, Priya, Sambac Jasmine, Tuscan Jasmine; **Sans. :** Ananya, Bhupadi, Devalata, Mallika, Sita; **Hindi:** Banmallika, Bel, Chamba, Moghra, Mogra, Motia, Mugra.

Distribution : Cultivated throughout India, upto 700m.

Part/s Used : Flowers.

Utilization : Flowers used for scenting hair oils.

Active Principles : Flowers yield 9'-deoxyjasminigenin, jasminin and 8,9-dihydrojasminin.

References : Ambasta (1986), Kirtikar and Basu (1984), Rastogi and Mehrotra (1993, 1995), Singh and Pandey (1998), Watt (1972).

296.*Jatropha gossypifolia* Linn.

Family : Euphorbiaceae.

English Name : **Eng. :** Bellyache Bush.

Distribution : Naturalized in many parts of India.

Part/s Used : Seeds.

Utilization : Dried seeds used as incense. Also, yield an essential oil.

Active Principles : These are β-amyrin, β-sitosterol, threonine, praline, glycine, alanine, valine, isoleucine, leucine, arginine, glutamic acid, xylose, arabinose, rhamnose and myristic, palmitic, stearic, aspartic, oleic and linoleic acids (seeds), alanine (28.1), glycine (19.71), valine (18.9), methionine (13.07), leucine (12.30), isoleucine (3.92), cystine (2.94) and arginine (0.98%) (seed essential oil).

References : Asolkar *et al.* (1992), Chopra *et al.* (1956), Jain and De (1966).

297.*Joannesia princeps* Vell.

Family : Euphorbiaceae.

Distribution : Grown in Indian gardens.

Part/s Used : Bark. Seeds.

Utilization : Bark oil smells of garlic. Seed oil, called Anda-assy, have somewhat agreeable odour.

References : Ambasta (1986), Asolkar *et al.* (1992), Rastogi and Mehrotra (1993).

298.*Juglans regia* L. var. ***kamaonia*** C.DC. (Pl. 8B)

Syn: J. arguta Wall.; *J. regia* non L.: Hook.f.

Family : Juglandaceae.

English Name : **Eng. :** The Himalayan Walnut Tree.

Distribution : Along the Himalaya from 1,170-2,300m.

Part/s Used : Nuts.

Utilization : Nuts deliciously flavoured, and employed in confectionery and cakes.

Active Principles : Juglone –5-OH-1,4-naphthaquinone, present in green walnut shells.

References : Asolkar *et al.* (1992), Bennett (1987), Harborne and Baxter (2001), Rastogi and Mehrotra (1995), Sahni (1998), Watt (1972).

299. *Juniperus chinensis* Linn.

Family : Cupressaceae.

English Name : **Eng. :** Chinese Juniper.

Distribution : Introduced into India.

Part/s Used : Wood.

Utilization : Wood used in cosmetics and as incense.

References : Ambasta (1986), Asolkar *et al.* (1992).

300. *Juniperus communis* Linn.

Syn: J.communis L. var. *saxatilis* Pallas; *J. nana* Willd.

Family : Cupressaceae.

English, Sanskrit and Hindi Names : **Eng. :** Common Juniper; **Sans.:** Aparajita, Habusha, Kaphaghni, Vapusha; **Hindi :** Aaraar, Abhal, Haubera.

Distribution : N.W. Himalaya from Kashmir to Kumaon westwards: 4,200-4,650m.

Part/s Used : Wood. Leaves. Fruits.

Utilization : Wood and leaves burnt as incense. Oil of Juniper and fruits used for flavouring gin, liquors and cordials.

Active Principles : Chief contents are cyclohexitol, terpinoid geijerone, umbelliferone, niloside, isoquercitrin (fruits), α-pinene (20%), limonene (9%), myrcene (9%) and borneol (8%) p-cymene, β-pinene, α-thujene, camphene and 1-terpinen-4-ol (oil).

References : Ambasta (1986), Asolkar *et al.* (1992), Chevallier (1996), Harborne and Baxter (2001), Greenish (1999), Kaul (1997), Kirtikar and Basu (1984), Purohit (2000), Rastogi and Mehrotra (1993), Rawat and Pangtey (1987), Watt (1972).

301. *Juniperus indica* Bertol.

Syn: J. excelsa Wall.; *J. pseudosabina* sensu Hook.f. (non Fisch and Mey); *J. wallichiana* Hook.f. and Thoms.

Family : Cupressaceae.

English and Hindi Names : **Eng. :** Black Juniper; **Hindi :** Bhil.

Distribution : Temperate Himalaya-Kashmir to Bhutan: 3,000-5,000m.

Part/s Used : Leaves. Twigs.

Utilization : Both leaves and twigs burnt as incense.

Active Principles : Amentoflavone, cupressuflavone, hinokiflavone and isocryptomerin isolated from leaves.

References : Kapur (1996a), Koelz (1979), Rastogi and Mehrotra (1995), Rawat and Pangtey (1987), Watt (1972).

302. ***Juniperus polycarpos*** C.Koch

Syn: J. excelsa Brand.; *J. gossainthaneana* Loddig.; *J. macropoda* auct. (non Boiss.)

Family : Cupressaceae.

English Names : **Eng. :** Himalayan Pine Cedar, Indian Juniper.

Distribution : Kajan valley, Kashmir, Lahaul extending to Kumaon: 1,500-4,300m.

Part/s Used : Fruit. Leaves. Twigs. Wood.

Utilization : Fruits, leaves, wood and twigs form part of the incense called 'dhoop'.

Active Principles : These are sugiol, 10-non-acosanol, β-sitosterol glucoside, flavoured junipodin, junipin (fruits), amentoflavone, hinokiflavone, isocryptomerialogwith quercetagetin-3-O-α-L-rhamnoside, quercetin-3-O-α-L-rhamnoside and kaempferol-3-O-β-D-glucoside, cupressuflavone and podocarpusflavone-A (leaves).

References : Asolkar *et al.* (1992), Kirtikar and Basu (1984), Rastogi and Mehrotra (1991, 1993), Sahni (1998), Sood *et al.* (2001), Watt (1972).

303. ***Juniperus recurva*** Buch.Ham.

Syn: J. densa Gord.; *J. lambertiana* Wall.; *J. religiosa* Royle; *J. rigida* Wall.; *J. recurva* var. *squamata* (Buch.-Ham. *ex* D.Don) Parl.; *J. squamata* Buch.Ham. *ex* D.Don.

Family : Cupressaceae.

English Name : **Eng. :** The Weeping Blue Juniper.

Distribution : Temperate and Alpine Himalaya: 2,500-5,000m.

Part/s Used : Wood. Leaves. Twigs.

Utilization : Heartwood fragrant. Wood, leaves and twigs used as incense.

Active Principles : Various constituents isolated are epicedranediol, 4-ketocedrol, 3β-hydroxycedrol, isocedrolic acid, β-chamigrenic acid (heart-wood), monoterpene hydrocarbons (12.0%), oxygen-ated sesquiterpenes (9.0%), oxygenated monoterpenes (5.0%) (essential oil of leaves and twigs), erpinene (24.01), camphene (17.68%) (leaf essential oil), biflavonoid cupressuflavone (leaves).

References : Ambasta (1986), Asolkar *et al.* (1992), Dam and Hajra (1981), Kirtikar and Basu (1984), Koelz (1979), Rastogi and Mehrotra (1991, 1993, 1995), Rawat and Pangtey (1987), Watt (1972).

304. ***Juniperus virginiana*** Linn.

Family : Cupressaceae.

English Names : **Eng. :** Pencil Cedar, Red Cedar.

Distribution : Introduced into Indian gardens as ornamental.

Part/s Used : Wood. Twigs. Fruits.

Utilization : Wood oil, (Cedarwood oil), used in perfumery whereas wood, twigs and fruits burnt as incense.

Active Principles : Cedarwood oil contains cedrene isomers (80%), cedrol (3-14%), cedrenol, thujopsene, β-elemene, caryophyllene, coparene and α-acaradiene.

References : Ambasta (1986), Asolkar *et al.* (1992), Harborne and Baxter (2001).

305.*Jurinea dolomiaea* Boiss.

Syn: J. macrocephala (Royle) C.B. Clarke

Family : Asteraceae.

Distribution : W. Himalaya (Kashmir to Kumaon): 3,650-4,600m.

Part/s Used : Roots.

Utilization : Aromatic roots used as commercial source of 'dhoop' (incense).

References : Bennett (1987), Kapur and Nanda (1996), Kaul (1997), Kirtikar and Basu (1984), Mittre (1981), Negi and Pant (1990), Rawat and Pangtey (1987), Sarin (1989), Singh (1999).

306.*Kaempferia angustifolia* Rosc.

Family : Zingiberaceae.

Hindi Names : **Hindi :** Kanjan-Bura, Mudu-Nirbisha.

Distribution : A native of Bengal.

Part/s Used : Roots.

Utilization : Roots used in perfumery.

References : Mehra *et al.* (1975), Watt (1972).

307.*Kaempferia galanga* Linn.

Syn: Alpinia sessilis Kon.

Family : Zingiberaceae.

English, Sanskrit and Hindi Names : **Eng. :** Black Thorn; **Sans. :** Chandramulika, Sugandhavacha; **Hindi :** Chandramula.

Distribution : Throughout the hotter parts of India; much cultivated in gardens.

Part/s Used : Rhizome. Roots. Tubers.

Utilization : Rhizome used as condiment and spices. Tubers employed in perfumery. Fragrant roots often attached to the necklaces of females for the sake of their perfume. Also, these are kept in clothes for the same purpose.

Active Principles : Rhizome contains deoxypodophyllotoxin, ethyl p-methoxy-trans-cinnamete, monoterpene ketone-car-3-en-5-one whereas roots possess Et-p-MeO-trans-cinnamate.

References : Ambasta (1986), Asolkar *et al.* (1992), Islam (1996), Rastogi and Mehrotra (1990, 1993, 1995), Rastogi and Mehrotra (1990, 1993, 1995), Tripathi and Goel (2001), Watt (1972).

308. *Kaempferia rotunda* Linn.

Family : Zingiberaceae.

Sanskrit and Hindi Names: **Sans.** : Bhuchampaca, Bhumichampa; **Hindi :** Bhuichampa.

Distribution : Throughout India mainly Assam, Meghalaya, Mizoram and Tripura.

Part/s Used : Rhizome. Leaves. Flowers.

Utilization : Tuberous rhizomes and young leaves used as a flavouring and flowers in perfumery.

Active Principles : Cineole and Methyl Chavicol (flowers).

References : Ambasta (1986), Chandra (1997), Kirtikar and Basu (1984), Mehtani and Ingle (1999), Watt (1972).

309. *Kedrotis foetidissima* (Jacq.) Cogn.

Syn: K. rostrata (Rottl.) Cogn.; *Rhynchocarpa foetida* Cl., non Schrad.; *Trichosanthes foetidissima* Jacq.

Family : Cucurbitaceae.

Distribution : W. Peninsula.

Part/s Used : Leaves.

Utilization : Leaves used as a flavouring.

References : Bennett (1987), Chandra (1997).

310. *Laggera aurita* (Willd.) Sch.-Bip. *ex* C.B. Clarke

Syn: Blumea aurita DC.; *B. guineensis* DC.; *Conyza aurita* Linn.

Family : Asteraceae.

Distribution : Punjab eastwards to W. Bengal, Orissa, Andhra Pradesh, Tamil Nadu, Kerela and Karnataka.

Part/s Used : Whole Plant.

Utilization : Plant strongly scented.

Active Principles : Plant contains laggerol, n-heptacosane, n-dotriacontane, δ-cadinene, α-cadinol, m-menth-6-en-8-ol and 2,3-dimethoxy-p-cymene.

References : Ambasta (1986), Rastogi and Mehrotra (1991), Watt (1972).

311. ***Lancea tibetica*** Hook.f. and Thoms.

Family : Scrophulariaceae.

Distribution : Himalaya (Kashmir to Sikkim): 3,300-4,800m.

Part/s Used : Leaves.

Utilization : Powdered leaves mixed with ghee to make incense.

Active Principles : Glucose, fructose, sucrose, β-sitosterol, sesamin and ursolic acid isolated from plant.

References : Rastogi and Mehrotra (1995), Singh *et al.* (1996).

312. ***Lantana camara*** Linn. var. ***aculeata*** (L.) Moldenke (Pl. 8C)

Syn: L. aculeata Linn.; *L. camara* auct., non Linn.

Family : Verbenaceae.

English Names : **Eng. :** Lantana Weed, Wild Sage.

Distribution : Naturalized in many parts of India.

Part/s Used : Leaves.

Utilization : Leaf oil with a pleasant lasting odour.

Active Principles : These are lantadene-D, camaroside (leaves) and sesquiterpenoids (oil).

References : Ambasta (1986), Bennett (1987), Kirtikar and Basu (1984), Mishra and Kumar (2000), Rastogi and Mehrotra (1995), Watt (1972).

313. ***Lantana rugosa*** Thunb.

Syn: L. salvifolia Jacq.

Family : Verbenaceae.

Utilization : Used for flavouring foods.

References : Ambasta (1986).

314. ***Laurentia papillosa*** (Forsk.) Grev.

Family : Rhodomelaceae.

Part/s Used : Algal Thallus.

Utilization : Used as a condiment.

Reference : Ambasta (1986).

315. ***Laurus nobilis*** Linn.

*Family :*Lauraceae.

English Names : **Eng. :** Sweet Bay, True Laurel.

Distribution : Grown in Indian gardens.

Part/s Used : Leaves.

Utilization : Fragrant laurel leaves employed as a condiment and for flavouring foods, confectionery and soups. Also, contain an essential oil (Laurel leaf oil) used for culinary purposes.

Active Principles : Chief constituents are costunolide, artemorin, verlotorin, santamarine, reynosin, 3-rutinoside, 3-glucoside, 3-galactoside, 3-rhamnoside, 30 arabinosides of kaempferol, quercetin (leaves), 1,8-cineole (26.7), eugenol (18.5), linalool (18.5), sabinene (11.8%) (leaf essential oil).

References : Ambasta (1986), Bhattacharjee (1998), Chevallier (1996), Greenish (1999), Harborne and Baxter (2001), Mehtani and Ingle (1999), Raju (2000), Rastogi and Mehrotra (1990, 1991, 1993, 1995).

316. ***Lavendula angustifolia*** Mill.

Syn: L. officinalis Chaix.; *L. spica* Linn.; *L. vera* DC.

Family : Lamiaceae.

English Names : **Eng. :** Common Lavender, True Lavender.

Distribution : Cultivated in Temperate regions of H.P.

Part/s Used : Flowers.

Utilization : Oil extracted from flowers and flowering tops called, Lavender oil, used in soap industry, high grade perfumery, production of Lavender water, and for flavouring pharmaceutical preparations.

Active Principles : The chief constituents of essential oil are linalool, linalyl acetate (30-60%), cineole (10%), pinene, limonene, nerol, borneol, umbellifeone and its methyl ether, monoterpenoids (I-VII), β-farnesene, β-bisabolene, caryophyllene, δ-cadinene, γ-cadinene and α-farnesene.

References : Ambasta (1986), Bhakuni (2000), Bhattacharjee (1998), Chevallier (1996), Harborne and Baxter (2001), Greenish (1999), Kumar and Singh (2000), Rastogi and Mehrotra (1990, 1991, 1995), Sharma and Sood (1997), Watt (1972).

317. ***Lavendula bipinnata*** Kuntze

Syn: L. burmann Benth.

Family : Lamiaceae.

Distribution : Chota Nagpur. Mount Abu. Jubbulpore. Konkan. Khandesh. Deccan.

Part/s Used : Flowers. Leaves.

Utilization : The essential oil from flowers has peppermint-like aroma and that from leaves characterized by lemongrass - like odour.

References : Ambasta (1986), Kirtikar and Basu (1984).

318. ***Lavendula hybrida*** Reverchon

Family : Lamiaceae.

English Name : **Eng. :** Lavandin.

Part/s Used : Flowers.

Utilization : The extracted oil from flowers combines the fragrance of true lavender and the camphoraceous harshness of spike lavender (*L. latifolia* Medic.).

Reference : Ambasta (1986).

319.*Lavendula latifolia* Medic.

Family : Lamiaceae.

English Name : **Eng.** : Spike Lavender.

Part/s Used : Flowers.

Utilization : Flowers yield an essential oil with harsh camphoraceous odour, used chiefly as a perfume.

Active Principles : Main constituents of the oil are linalool 40-50%, 1,8-cineole 20-30%, camphor 10-20% and linalyl acetate 1%.

References : Ambasta (1986), Chopra *et al.* (1960), Harborne and Baxter (2001).

320.*Lavendula stoechas* Linn.

Family : Lamiaceae.

English and Hindi Names : **Eng.** : French Lavender; **Hindi** : Dharu.

Distribution : Found throughout the Mediterranean region to Asia minor.

Part/s Used : Flowers.

Utilization : Flowers camphoraceous and used in perfumes.

Active Principles : Essential oil extracted from flowers contains camphor (47.06), borneol (5.37), linalyl acetate, geranial (2.98 each), cineole (2.51), caryophyllene (2.37), α-terpineol (1.09), neral and pulegone (0.85), citronellol and fenchol (0.79 each), camphene (0.50), β-pinene and fenchone (0.45 each),α-pinene (0.42), β-terpineol (0.39), menthol (0.35), α-thujone (0.33%) and isoborneol.

References : Ambasta (1986), Chopra *et al.* (1980), Rastogi and Mehrotra (1990, 1991, 1995), Watt (1972).

321.*Lawsonia alba* Lamk.

Syn.: *L. inermis* Linn.; *L. purpurea* Lamk.; *L. spinosa* Linn.

Family : Lythraceae.

English, Sanskrit and Hindi Names : **Eng.** : Camphire, Cypress Shrub, Egyptian Privet, Henna Plant, Jamaica Mignonette, Mindie; **Sans.** : Dvivranta, Kokadanta, Mendika, Ragangi, Raktagarbha, Timira; **Hindi** : Hena, Mehndi.

Distribution : Throughout India, chiefly in drier parts of Peninsula.

Part/s Used : Flowers.

Utilization : Fragrant flowers valued in perfumery.

Active Principles : Henna contains coumarins, naphthaquinones, flavonoids, sterols and tannins. Flowers afford lower amounts of mannitol whereas its essential oil (0.02% yield) contains β-ionone.

References : Bennett (1987), Chevallier (1996), Chopra *et al.* (1980), Kirtikar and Basu (1984), Rastogi and Mehrotra (1995), Singh and Pandey (1998), Vihari (1995), Watt (1972).

322.*Lepidium sativum* Linn.

Family : Brassicaceae.

English, Sanskrit and Hindi Names : **Eng. :** Common Cress, Garden Cress; **Sans. :** Ashalika, Bhadra, Candrika, Chandrashura, Chandrika, Nandini, Suvasura; **Hindi :** Chansur, Halim, Hurf, Malavan.

Distribution : Cultivated throughout India.

Part/s Used : Seeds.

Utilization : Seeds used in the flavouring. Also, yield a volatile aromatic oil.

Active Principles : Seeds rich in lepidine along with sinapic acid ethyl ester, N, N'-dibenzylthiourea and N,N'-dibenzylurea.

References : Anonymous (1999), Kirtikar and Basu (1984), Negi *et al.* (1993), Rastogi and Mehrotra (1990, 1995), Watt (1972).

323.*Leucas aspera* Spreng.

Syn: L. dimidiata Benth.; *L. plukenetii* Benth.; *Phlomis aspera* Willd.; *P. esculenta* Roxb.; *P. obliqua* Ham.; *P. plukenetii* Roth.

Family : Lamiaceae.

Hindi Name : **Hindi :** Chota Halkusa.

Distribution : Throughout the plains of India extending from Sikkim and Bihar westwards to the Punjab, and southward to Cape comorin.

Part/s Used : Whole Plant.

Utilization : Fragrant plant employed as a pot-herb in times of scarcity.

References : Ambasta (1986), Kirtikar and Basu (1984), Watt (1972).

324.*Leucas lavandulaefolia* Rees.

Syn: L. indica (L.) R.Br. *ex* Vatke; *L. linifolia* (Roth) Spreng.; *Leonurus indicus* Burm. var. *decipiens* Sm.; *Phlomis linifolia* Roth.; *P. zeylanica* Roxb.

Family : Lamiaceae.

Sanskrit and Hindi Names : **Sans. :** Dronapushpi, Kumbhi, Rudrapushpa; **Hindi :** Guma, Halkusa, Kumbha.

Distribution : More or less throughout India.

Part/s Used : Whole Plant.

Utilization : Plant aromatic, used as a flavouring.

Active Principles : Acacetin and chrysoeriol isolated from aerial parts.

References : Ambasta (1986), Kirtikar and Basu (1984), Rastogi and Mehrotra (1995), Watt (1972).

325. ***Leucas martinicensis*** R.Br.

Family : Lamiaceae.

Distribution : All over S. India.

Part/s Used : Leaves.

Utilization : Plant has mint-like fragrance.

References : Ambasta (1986), Chopra *et al.* (1956).

326. ***Leucas zeylanica*** R.Br.

Syn: L.involucrata Benth.; *Leonurus marrubiastrum* Burm.; *Phlomis zeylanica* Linn.

Family : Lamiaceae.

Distribution : Assam, W. Ghats. Travancore.

Part/s Used : Leaves.

Utilization : Leaves used for flavouring, yields an essential oil.

References : Ambasta (1986), Kirtikar and Basu (1984), Watt (1972).

327. ***Lilium candidum*** Linn.

Family : Liliaceae.

English Names : **Eng.** : Annunciation Lily, Madona Lily.

Distribution : Grown as ornamental in India.

Part/s Used : Flowers.

Utilization : Fragrant flowers used in high grade perfumery.

Active Principles : Lilyn, 2-phenylethylpalmitate and 2-phenylethyl-α-L-arabinosyl (1 to 6)-β-D-glucoside isolated from petals.

References : Ambasta (1986), Rastogi and Mehrotra (1991,1995).

328. ***Lilium giganteum*** Wall.

Syn: L.cordifolium Don.

Family : Liliaceae.

Distribution : Temperate Himalaya (Kumaon and Garhwal to Sikkim): 1,700-3,500 m, Khasia Hills.

Part/s Used : Flowers.

Utilization : Flowers fragrant.

References : Kirtikar and Basu (1984), Raju (2000), Watt (1972).

329. *Limnophila indica* (Linn.) Druce

Syn: L. gratioloides R.Br.; *L. racemosa* Benth.

Family : Scrophulariaceae.

Sanskrit and Hindi Names: **Sans. :** Ambuja; **Hindi :** Kuttra.

Distribution : Throughout India in wet places.

Part/s Used : Leaves.

Utilization : Plant odour akin to that of camphor or oil of lemons.

References : Ambasta (1986), Bennett (1987), Chopra *et al.* (1956).

330. *Limnophila rugosa* (Roth.) Merrill

Syn: L. roxburghii G.Don

Family : Scrophulariaceae.

Distribution : Throughout India.

Part/s Used : Leaves.

Utilization : Leaves used for flavouring and perfuming hair. Leaves also yield an essential oil akin to *Ocimum basilicum* and used for flavouring foods.

Active Principles : Essential oil of leaves yields p-methoxy benzoic acid, anisaldehyde, anethole, estragole and unidentified substituted phenyl propane derivatives .

References : Ambasta (1986), Bennett (1987), Chopra *et al.* (1980), Rastogi and Mehrotra (1991,1995).

331. *Lindera neesiana* Benth.

Syn: Aperula neesiana Blume; *Benzoin neesianum* Nees.; *Laurus macrophylla* Don.; *Tetranthera neesiana* Wall.; *T. prunifolia* Wall.

Family : Lauraceae.

Distribution : Temperate Himalaya. Khasia and Naga Hills in Assam.

Part/s Used : Whole Plant.

Utilization : Aromatic tree substituted for true sassafras from *Cinnamomum glanduliferum* Meissn.

Active Principles : Seeds contain methyl laurate (75.2), methyl caprate (13.3), methyl oleate (5.4), methyl myristate (2.4) and methyl palmitate (0.5%).

References : Ambasta (1986), Chopra *et al.* (1980), Kirtikar and Basu (1984), Rastogi and Mehrotra (1995), Watt (1972).

332. *Lindera pulcherrima* Benth.

Syn: Daphnidium pulcherrimum Nees.

Family : Lauraceae.

Distribution : Temperate Himalaya (1,350-3,000 m), Khasia mountains (1,700-2,000 m).

Part/s Used : Bark.

Utilization : Bark used as spice.

References : Saklani and Jain (1994), Watt (1972).

333. *Liquidambar orientalis* Mill.

Syn : *L.imberbe* Aiton

Family : Altingiaceae.

English, Sanskrit and Hindi Names: **Eng.:** Liquidambar, Liquid Storax, Oriental Sweet Gum, Rose Malloes; **Sans.:** Silhaka; **Hindi:** Silaras.

Distribution : Forming forests in the extreme S.W. part of Asia minor.

Part/s Used : Balsam obtained from trunk (Storax).

Utilization : Used for scenting soaps and in oriental perfumes.

Active Principles : Oily liquid contains styrol (phenylethylene, an aromatic liquid), ethyl cinnamate, phenyl propyl cinnamate, cinnamyl cinnamate (=styracin), vanillin and free cinnamic acid.

References : Ambasta (1986), Chevallier (1996), Greenish (1999), Watt (1972).

334. *Litsea cubeba* (Lour.) Pers.

Syn : *Laurus cubeba* Lour.; *Litsea citrata* Blume; *Litsea kingii* Hook.f.

Family : Lauraceae.

Part/s Used : Flowers. Fruits.

Utilization : Oil from flowers used in perfumery. Fruits lemon scented and used as spice for flavouring chutney. Also yield an oil, 'Litsea cubeba oil', used as a fragrance component in deodorants, toiletries, perfumes, detergents, air fresheners, and fruit products.

Active Principles : Essential oil from fruits contain (+) limonene (6.0), (-) sabinene (7.0), methylheptenone (2.5), citronellal (1.0), citrals (80%), furfural, formaldehyde, acetone, butyraldehyde, isovaleraldehyde and methylheptenone.

References : Ambasta (1986), Bennett (1987), Chopra *et al.* (1980), Harborne and Baxter (2001), Kulkarni and Kumbhojkar (1993), Lalramnghinglova (1996), Rastogi and Mehrotra (1990).

335. *Lobaria pulmonaria* (Linn.) Hoffm.

Family : Stictaceae.

Part/s Used : Lichen Thallus.

Utilization : Used in perfumery.

Active Principles : Contains a variety of plant acids (including stictic and sticinic acid), fatty acids, mucilage and tannins.

References : Ambasta (1986), Chevallier (1996).

336.*Macadamia ternifolia* F.Muell.

Family : Proteaceae.

Distribution : Cultivated in Nilgiris, Chennai, Mysore and U.P.

Part/s Used : Kernels.

Utilization : Kernels resemble hazel nuts and almonds in flavour and used in confectionery. Also yield an oil, used as salad oil and for high grade soaps.

References : Ambasta (1986), Chopra *et al.* (1980).

337.*Madhuca indica* J.F. Gmel.

Syn : *Bassia latifolia* Roxb.; *B.longifolia* Koenig. in L.; *M.latifolia* (Roxb.) Macbr.; *M.longifolia* (Koenig) *Mac Bride* var. *latifolia* Cheval.

Family : Sapotaceae.

English, Sanskrit and Hindi Names: **Eng.** : Illipe, Mahawash Tree, Mahua, Mowra, The Indian Butter Tree; **Sans.** : Gudapuspa; **Hindi** : Mahua, Mauwa, Mohwa.

Distribution : Himalayan foothills. Southwards to C. India.

Part/s Used : Flowers.

Utilization : Flowers scented.

Active Principles : Polysaccharide PS-AI and PS-AII isolated from flowers.

References : Ambasta (1986), Anonymous (1999), Bennett (1987), Chevallier (1996), Rastogi and Mehrotra (1991,1993), Sahni (1998), Watt (1972).

338.*Magnolia denudata* Desr.

Syn : *M.conspicua* Salisb.

Family : Magnoliaceae.

Distribution : Reported to be grown in Indian gardens.

Part/s Used : Buds.

Utilization : Buds used for seasoning rice.

References : Ambasta (1986), Rastogi and Mehrotra (1993).

339.*Mahonia acanthifolia* G.Don

Syn : *Berberis acanthifolia* Wall.

Family : Berberidaceae.

Distribution : Himalaya - Kumaon to Bhutan and Arunachal Pradesh: 2,400-2,900m.

Part/s Used : Flowers.

Utilization : Flowers scented.

References : Negi and Gaur (1994), Rastogi and Mehrotra (1991).

340.*Majorana hortensis* Moench.

Syn : Origanum majorana Linn.

Family : Lamiaceae.

English and Hindi Names: **Eng. :** Sweet Marjoram; **Hindi :** Murwa.

Distribution : Commonly grown in Indian gardens.

Part/s Used : Flowering Tops. Leaves . Seeds.

Utilization : Leaves used as condiment and also for flavouring processed meat products, soups, salads, vinegar, dressings for stews, vegetables, cheese and poultry dishes. Leaves and flowering tops yield sweet Marjoram oil, used in high grade flavours, perfumes, soap and liqueur industries. Aromatic seeds used in confectionery.

Active Principles : These are majoranin (leaves), eugenol (26.0) and p-cymene (14.0%) (leaf oil), eugenol (6.7%), chavicol (4.6%) and α-linalool (30.6%) (leaves and flowering stalks).

References : Ambasta (1986), Chopra *et al.* (1980), Mehtani and Ingle (1999), Rastogi and Mehrotra (1991, 1993, 1995).

341.*Malpighia punicifolia* Linn.

Family : Malpighiaceae.

English Name : **Eng. :** West Indian Cherry.

Distribution : Cultivated in gardens of Karnataka and Tamil Nadu.

Part/s Used : Fruits.

Utilization : Fruits used in flavouring jams, squashes, preserves and sauces.

Active Principles : Fruits contain 3-methyl-3-buten-1-ol as major volatile constituent.

References : Ambasta (1986), Mehtani and Ingle (1999), Rastogi and Mehrotra (1995).

342.*Malus baccata* (Linn.) Borkh.

Syn : M.baccata (L.) Borkh. var. *himalaica* (Maxim.) Schneid.; *Pyrus baccata* Linn. var. *himalaica* (Maxim.) Schneid.

Family : Rosaceae.

English and Hindi Names: **Eng. :** Himalayan Crab Apple, Siberian Crab Apple; **Hindi :** Ban Mehal, Gwalam.

Distribution : Temperate Himalaya.

Part/s Used : Fruits.

Utilization : Fruits have true apple flavour.

Active Principles : β-sitosterol, campesterol and ursolic acid isolated from fruits.

References : Ambasta (1986), Negi and Gaur (1994), Rastogi and Mehrotra (1995), Watt (1995).

343. *Mammea americana* Linn.

Family : Clusiaceae.

English Name : **Eng. :** Mammey Apple.

Distribution : Cultivated in India.

Part / s Used : Fruits. Flowers.

Utilization : An aromatic liquor, called Eau-de-Creole, is distilled from the flowers for use as a flavouring.

Active Principles : Mammeisin isolated from fruit pulp whereas fruit peels afford 4-phenyl-5,7-dihydroxy-6-isovaleryl-8-(3-methyl-2-buten -1-yl)-coumarin (mammeisin).

References : Ambasta (1986), Rastogi and Mehrotra (1990).

344. *Mammea siamensis* (Miq.) T. Anders

Syn : Calyssacion siamense Miq.; *Ochrocarpus siamensis* T. Anders.

Family : Clusiaceae.

Distribution : Met with in Lushai hills in Nagaland.

Part / s Used : Flowers.

Utilization : Flowers yield an essential oil with the aroma of violets.

Active Principles : Two phenylcoumarins – 6-butyryl-5-hydroxy-4-phenyl-seselin (I) and 6-butyryl-5,7-dihydroxy-8-(3,3-dimethyl-allyl)-4-phenylcoumarin (II) isolated from flowers.

References : Ambasta (1986), Bennett (1987), Rastogi and Mehrotra (1993).

345. *Mangifera indica* Linn.

Syn : M. domestica Gaertn.

Family : Anacardiaceae.

English, Sanskrit and Hindi Names: **Eng. :** Cuckoo's Joy, Cupid's favourite, Mango Tree, Spring Tree; **Sans.:** Amra, Bhramarapriya, Chutaka, Kamanga, Kamaphala, Manjari, Priyambu, Sripriya; **Hindi :** Am, Amb.

Distribution : Cultivated in the subcontinent, for atleast 4,000 years. Wild in the foothills of the E. Himalaya and the W. Ghats.

Part / s Used : Fruits.

Utilization : Powdered dried unripe fruits 'Amchur' used as a spice for its acidic taste and in culinary preparations.

Active Principles : Flavour principle is due to car-3-ene and dimethylstyrene.

References : Ambasta (1986), Harborne and Baxter (2001), Kirtikar and Basu (1984), Raju (2000), Rastogi and Mehrotra (1993, 1995), Sahni (1998), Singh and Pandey (1998), Watt (1972).

346. ***Matricaria chamomilla*** **Linn.**

Syn : M. suaveolens Linn.

Family : Asteraceae.

English Names: **Eng. :** German Chamomile, Persian Chamomile, Wild Chamomile.

Distribution : Introduced in India in Lucknow and Jammu.

Part/s Used : Flowers.

Utilization : Essential oil, called Oil of German Chamomile, used in high grade perfumes and for scenting chamomile shampoo powders, tobacco, alcoholic and non-alcoholic beverages, ice-creams, baked goods and chewing gums.

Active Principles: Essential oil extracted from flowers possesses bisabolol oxide, borneol, caryophyllene, chamazulene, 1,8-cineole, farnesene, linalool, myrcene, pulegone and α-terpineol.

References : Ambasta (1986), Bhattacharjee (1998), Kaul (1997), Kirtikar and Basu (1984), Kumar and Singh (2000), Rastogi and Mehrotra (1990, 1991, 1993, 1995), Sharma and Sood (1997), Watt (1972).

347. ***Melissa officinalis*** **Linn.**

Family : Lamiaceae.

English Name : **Eng. :** Lemon Balm.

Distribution : Grows throughout the world.

Part/s Used : Whole Herb.

Utilization : Herbs and the extracted oil of balm from it is used for flavouring soups, salads, liquors and in perfumery.

Active Principles : Oil contains citral, caryophyllene oxide, linalool and citronellal.

References : Ambasta (1986), Chevallier (1996).

348. ***Mentha aquatica*** **Linn.**

Family : Lamiaceae.

English Names : **Eng. :** Marsh Mint, Water Mint.

Distribution : Cultivated in Indian gardens.

Part/s Used : Leaves.

Utilization : Leaves added to food and liquors for flavouring. Also, yields an essential oil.

Active Principles : Menthofuran, menthone, menthol, menthyl acetate and viridifloral isolated from essential oil.

References : Ambasta (1986), Chopra *et al.* (1980), Mehtani and Ingle (1999), Rastogi and Mehrotra (1991, 1995).

349.*Mentha arvensis* Linn. subsp. ***haplocalyx*** Briq. var. ***piperascens*** Holmes (Pl. 8D)

Family : Lamiaceae.

English and Hindi Names : **Eng. :** Corn Mint, Japanese Mint, Marsh Mint; **Hindi :** Pudinah.

Distribution : W. Himalaya: 1,500-3,000m.

Part/s Used : Leaves.

Utilization: Oil extracted from leaves called Japanese mint oil or Japanese peppermint oil, used as a substitute for true peppermint oil (*M. piperita* Linn.) and used to flavour confectionery items like candy, chocolates, chewing gum and cosmetic industries.

Active Principles : Chief ingredients of the oil are 1-menthol (73.0%), d-menthone, menthyl acetate, carvomenthone, limonene, piperitone, β-phellandrene, neomenthol and isomenthone.

References : Ambasta (1986), Bhakuni (2000), Jain (1968), Kirtikar and Basu (1984), Kumar and Singh (2000), Mehtani and Ingle (1999), Misra and Kumar (2000), Rastogi and Mehrotra (1991, 1993).

350.*Mentha cardiaca* Gerard *ex* Baker

Family : Lamiaceae.

Part/s Used : Whole Herb.

Utilization : Mint oil used for flavouring chewing gums, tooth pastes, confectionery and pharmaceutical preparations.

References : Ambasta (1986), Bhakuni (2000), Kumar and Singh (2000).

351.*Mentha citrata* Ehrh. Beitr.

Family : Lamiaceae.

English Name : **Eng. :** Bergamot mint.

Part/s Used : Whole Herb.

Utilization : Aromatic oil called bergamot mint, extracted from plant finds use in perfumery.

Active Principles : Oil yields linlool and linalyl acetate.

References : Bhakuni (2000), Kumar and Singh (2000), Misra and Kumar, (2000).

352.*Mentha longifolia* (Linn.) Nath.

Family : Lamiaceae.

English, Sanskrit and Hindi Names: **Eng. :** Horsemint; **Sans. :** Ajirnahara, Pudina, Rochani, Sugandhipatra; **Hindi :** Jangli Pudina, Podina.

Distribution : Temperate W. Himalaya: 1,350-4,000m; frequently cultivated in the plains of India.

Part/s Used : Leaves.

Utilization : Leaves and flowering tops yield an essential oil used for flavouring chutneys and confectionery.

Active Principles : Leaf oil contains eucalyptol, piperitenone oxide, piperitone oxide, α-terpineol, (-) carvone (76.9), trans-carveol (8.8), *cis*-carveol (2.2), trans-dihydrocarveol (5.4), *cis*-dihydro-carveol (3.4), limonene (2.1), 1,8-cineole (1.2%), humulene and caryophyllene.

References : Ambasta (1986), Kaul (1997), Kirtikar and Basu (1984), Mehtani and Ingle (1999), Purohit (2000), Rastogi and Mehrotra (1991, 1993, 1995), Watt (1972).

353.*Mentha piperita* Linn. emend. Huds.

Family : Lamiaceae.

English and Hindi Names : **Eng.** : Brandy Mint, Peppermint; **Hindi** : Gamathi Phudina, Paparaminta.

Distribution : Commercially cultivated in Kashmir, Dehradun, Nilgiris and Mysore.

Part/s Used : Leaves.

Utilization : Leaves used as a flavouring agent for the preparation of mint teas, mint sauces and mint salads. Peppermint oil extracted from leaves used to flavour non-alcoholic beverages, ice-creams, candies, baked goods, gelatins and puddings, chewing gum and some meat preparations.

Active Principles : Leaves contained menthoside and isorhoifolin, whereas thymol, carvacrol, eugenol, menthone, isomenthone, carvone, piperitone, β-pinene, cadinene, menthofuran, dipentene, menthyl acetate, 1,8-cineole, neomenthol, linalool, isomenthol, menthol, keto-alcohol (I), (-) mintlactone and (+) isomintlactone isolated from essential oil.

References : Ambasta (1986), Bhakuni (2000), Bhattacharjee (1998), Chevallier (1996), Kirtikar and Basu (1984), Mehtani and Ingle (1999), Misra and Kumar (2000), Purohit (2000), Raju (2000), Rastogi and Mehrotra (1990, 1991, 1993, 1995), Watt (1972).

354.*Mentha pulegium* Linn.

Family : Lamiaceae.

English Names: **Eng.** : English or European Pennyroyal, Pudding Grass.

Distribution : Introduced into India.

Part/s Used : Leaves.

Utilization : Leaves used for flavouring baked goods, beverages, and puddings. Yields an essential oil, Pennyroyal Oil used for scenting soaps.

Active Principles : Pennyroyal's essential oil contains pulegone, isopulegone, pulegol, menthone, cadinene, piperitenol, isomenthone, piperetyl acetate and other terpenoids.

References : Ambasta (1986), Chevallier (1996), Mehtani and Ingle (1999), Rastogi and Mehrotra (1990, 1991).

355.*Mentha rotundifolia* (Linn.) Huds.

Family : Lamiaceae.

English Names: **Eng. :** Apple-Mint, Round Leaved Mint.

Distribution : Cultivated to a small extent in Indian gardens.

Part/s Used : Herb.

Utilization : Herb used in the flavouring. Yields an essential oil.

Active Principles : Carvone, pulegone, menthone, isomenthone, cineole, limonene, valeric acid, o-cresol, p-cresol and isopiperitenone are major constituents of essential oil.

References : Ambasta (1986), Rastogi and Mehrotra (1990, 1991, 1993).

356.*Mentha spicata* Linn. emend. Nathh.

Syn : M.spicata var. *viridis* Linn.; *M.viridis* Linn.

Family : Lamiaceae.

English and Hindi Names : **Eng. :** Browm Mint, Garden Mint, Lamb Mint, Mackerel Mint, Spearmint; **Hindi :** Pahari Pudina, Pudina.

Distribution : Commonly cultivated in native gardens all over the plains of India.

Part/s Used : Leaves. Flowering Tops.

Utilization : Leaves used as condiment and for flavouring culinary preparations, vinegar, jellies and iced drinks. Essential oil, called Spearmint oil extracted from flowering tops also used for flavouring non-alcoholic beverages, baked goods, gelatin and puddings, ice-creams, chewing gums, candy, tooth pastes and confectionery.

Active Principles : Major components of the Spearmint oil are carvone (66.9), limonene (10.84%), phellandrene, dihydro-carvone pinen, phellandrene, 1-limonene, octyl alcohol and dihydrocarveol.

References : Ambasta (1986), Bhakuni (2000), Bhattacharjee (1998), Harborne and Baxter (2001), Kaul (1997), Kirtikar and Basu (1984), Kumar and Singh (2000), Mehtani and Ingle (1999), Purohit (2000), Rastogi and Mehrotra (1991, 1993, 1995), Singh and Pandey (1998), Watt (1972).

357. *Meriandra benghalensis* Benth.

Syn : Salvia bengalensis Benth.; *S.dianthera* Roth.

Family : Lamiaceae.

English and Hindi Names : **Eng. :** Bengal Sage; **Hindi :** Kafurkapat.

Distribution : Cultivated in India especially in Bengal.

Part/s Used : Leaves.

Utilization : Leaves employed as a condiment. Herb yields an essential oil, called oil of Meriandra, used for the same purpose as the sage oil.

Active Principles : Essential oil yields camphor 5.9-70.5% and cineole 9.7-24.2%.

References : Ambasta (1986), Chopra *et al.* (1980), Kirtikar and Basu (1984), Rastogi and Mehrotra (1993, 1995), Watt (1972).

358. *Merremia dissecta* (Jacq.) Hallier.f.

Syn : Convolvulus dissectus Jacq.; *Ipomoea sinuate* Ort.

Family : Convolvulaceae.

Part/s Used : Seeds.

Utilization : Seeds possess odour akin to oil of bitter almonds and used in the preparation of a liquor.

Active Principles : Amygdalin-6"-(4-hydroxy) benzoate and amyg-dalin-6"-(4-hydroxy)-(E) cinnamate, amygdalin isolated from seeds.

References : Ambasta (1986), Bennett (1987), Rastogi and Mehrotra (1995).

359. *Messerschmidia argentea* (Linn.f.) Johnston

Syn : Tournefortia argentea Linn.f.

Family : Boraginaceae.

Part/s Used : Leaves.

Utilization : Leaves eaten raw for their parsley like flavour.

Reference : Ambasta (1986).

360. *Mesua ferrea* Linn.

Syn : M. coromandeliana Wight; *M. ferrea* ssp. *thwaitesii* Vesque in DC.; *M. nagana* Gard.; *M. pedunculata* Wight; *M. pulchella* Planch. and Trian.; *M. roxburghii* Wight; *M. salicina* Planch. and Trian.; *M. sclerophylla* Thwaites.; *M. speciosa* Chois in DC.; *M. thwaitesii* Planch. and Trian.; *M. walkeriana* Planch. and Trian.

Family : Clusiaceae.

English, Sanskrit and Hindi Names: **Eng. :** Ceylon Iron-Wood, Cobras Saffron, Iron wood of Assam; **Sans. :** Hema, Kesara, Nagakeshara, Pinjara, Suvarna; **Hindi :** Naghas, Nagkesar.

Distribution : Evergreen and Semi-evergreen forests of N.E. India. W. Ghats. Andamans.

Part/s Used : Flowers. Stamens.

Utilization : Fragrant dried flowers used as a fragrance component in perfumed ointments. Stamens fragrant and resembling a rose, and used for stuffing pillows.

Active Principles : Stamens contain mesuaferrone A, mesuaferone B and mesuaferrol.

References : Ambasta (1986), Anonymous (1999), Bennett (1987), Bhattacharjee (1998), Kirtikar and Basu (1984), Rastogi and Mehrotra (1991, 1995), Sahni (1998), Sinha and Sinha (2001), Watt (1972).

361.*Michelia champaca* Linn.

Syn : *M. aurantiaca* Wall.; *M. doltsopa* Buch.-Ham *ex* DC.; *M. excelsa* (Wall.) Blume; *M. rheedi* Wight.; *M. rufinervis* DC.

Family : Magnoliaceae.

English, Sanskrit and Hindi Names: **Eng.** : Champak, Golden Champa, Yellow Champa; **Sans.** : Anjana, Champaka, Deepapushpa, Gandhaphali, Hemapushpa, Kusuma, Surabhi; **Hindi** : Champa, Champaca.

Distribution : E. Sub-Himalaya. W. Ghats: 500-1,500m. S. India. Nilgiris.

Part/s Used : Flowers. Leaves. Wood.

Utilization : Highly fragrant flowers source of sweet-smelling Champa Oil, much valued in preparation of 'attars' and perfumed hair oils. Leaves yield an essential oil with odour akin to that of Basil (*O. basilicum* Linn.). Camphor extracted from the wood used in making agarbatties.

Active Principles : Leaves contain lanuginosine, michelanugine, liriodenine and sitosterol.

References : Ambasta (1986), Bhattacharjee (1998), Kirtikar and Basu (1984), Mittre (1981), Raju (2000), Rastogi and Mehrotra (1991), Sahni (1998), Vihari (1986), Watt (1972).

362.*Michelia figo* (Lour.) Spreng.

Syn : *M. fuscata* Blume; *Magnolia fuscata* Andr.

Family : Magnoliaceae.

Distribution : Grown in Indian gardens.

Part/s Used : Flowers.

Utilization : Flowers used for scenting tea and in perfumery. Also yield an essential oil.

Active Principles : Flowers possess magnolamine, dehydrolan-uginolide, lipiferolide, syringaresinol, deacetyllanuginolide and michefusc-alide.

References : Ambasta (1986), Rastogi and Mehrotra (1993).

363.*Michelia kisopa* Buch.-Ham. *ex* DC.

Family : Magnoliaceae.

Part/s Used : Flowers.

Utilization : Flowers sweetly scented.

Reference : Ambasta (1986).

364.*Micromeria capitellata* Benth.

Family : Lamiaceae.

Distribution : Upper Gangetic Plain, Chota Nagpur, Bihar, W.Ghats, N. Circars, Nilgiris, Parisnath and the W. Himalaya.

Part/s Used : Whole Plant.

Utilization : Strongly aromatic herb used for the same purposes as mentha (*Mentha piperita* Linn.). Also, yield an oil.

Active Principles : Oil yields pulegone (80.0%).

References : Ambasta (1986), Kirtikar and Basu (1984), Rastogi and Mehrotra (1995), Watt (1972).

365.*Microtoena insuavis* (Hance) Prain *ex* Dunn

Syn : Gomphostemma insuavis Hance; *M. cymosa* Prain; *Plectranthus patchouli* C.B. Clarke.

Family : Lamiaceae.

Distribution : Tripura. Assam. Manipur. Khasia. Nagaland.

Part/s Used : Whole Herb.

Utilization : The herb smells strongly of Patchouli obtained from *Pogostemon* spp.

References : Ambasta (1986), Bennett (1987), Chopra *et al.* (1980).

366.*Millettia extensa* (Benth.) Baker

Syn : M. auriculata Baker *ex* Brandis; *Otosema extensa* Benth. in Miq.

Family : Papilionaceae.

Distribution : Outer Himalayas from the Satluj eastwards to Sikkim, upto 1,200m.

Part/s Used : Flowers.

Utilization : Flowers fragrant.

Active Principles : Auriculatin isolated from flowers.

References : Bennett (1998), Chopra *et al.* (1956), Rastogi and Mehrotra (1990), Singh and Pandey (1998).

367. ***Mimusops elengi*** Linn.

Family : Sapotaceae.

Sanskrit and Hindi Names: **Sans.** : Anangaka, Bakula, Chirapushpa, Dhanvi, Kantha, Mukula, Visharada; **Hindi** : Bakul, Bolsari, Maulsarau, Maulser, Mulsari.

Distribution : The Indian Peninsula along the W. Ghats. Andamans.

Part / s Used : Flowers.

Utilization : Fragrant flowers used in perfumery.

Active Principles : Flowers yield D-mannitol and β-sitosterol

References : Ambasta (1986), Bhattacharjee (1998), Kirtikar and Basu (1984), Rastogi and Mehrotra (1990, 1991), Sahni (1998), Seth *et al.* (1962), Watt (1972).

368. ***Momordica balsamina*** Linn.

Family : Cucurbitaceae.

English and Hindi Names: **Eng.** : Balsam Apple, Balsamina; **Hindi** : Mokha.

Distribution : N. and W. India, upto 300m.

Part / s Used : Fruits.

Utilization : Tender fruits used for flavouring food dishes.

Active Principles : Fruits rich in momordicin.

References : Ambasta (1986), Chandra (1997), Chopra *et al.* (1956), Kirtikar and Basu (1984), Mehtani and Ingle (1999), Watt (1972).

369. ***Momordica charantia*** Linn.

Syn : M. muricata DC.; *M. senegalensis* lamk.

Family : Cucurbitaceae.

English, Sanskrit and Hindi Names: **Eng.** : Bitter Gourd, Carilla Fruit; **Sans.** : Ambuvallika, Brihadvalli, Chiripatra, Karavalli, Karavella, Kathilla, Varivalli; **Hindi** : Karela, Karola, Kareli.

Distribution : Cultivated throughout India.

Part / s Used : Leaves. Fruit.

Utilization : Fruit and leaves used for flavouring food dishes.

Active Principles : These are 3 cucurbitane triterpenoids (I, II and III), momordicines I and II (leaves), charantin-β, 5,25 – diene- 3β-O-glucoside, a mixture of two acylglycosylsterols-3-O-[6'-O-palmitoyl-β-D-glucosyl], stigmasta-5,25 (27)-diene, benzyl alcohol, myrtenol, cis-3-hexenol, trans-2-hexenal, 1-penten-3-ol and cis-2-penten-1-ol (fruits).

References : Anonymous (1999), Chevallier (1996), Kirtikar and Basu (1984), Mehtani and Ingle (1999), Rastogi and Mehrotra (1990, 1993, 1995), Watt (1972).

370.*Monstera deliciosa* Liebm.

Syn : Philodendron pertusum Kunth and Bouche

Family : Araceae.

English Name : **Eng. :** Ceriman.

Distribution : Introduced into Indian gardens.

Part/s Used : Fruits.

Utilization : The fruiting spikes have mixed flavour of banana and pineapple and considered a delicacy.

References : Ambasta (1986), Bajpayee and Dixit (1996), Rastogi and Mehrotra (1995).

371.*Morina coulteriana* Royle

Family : Dipsacaceae.

Distribution : Sub-Alpine Himalaya : Kashmir to Garhwal.

Part/s Used : Roots. Flowers.

Utilization : Roots and flowers used as an incense.

References : Ambasta (1996), Kaul (1997), Sood *et al.* (2001), Watt (1972).

372.*Morina longifolia* Wall.

Family : Dipsacaceae.

Distribution : Temperate and Alpine Himalaya from Kashmir to Bhutan :2,400-4,200m.

Part/s Used : Whole Plant.

Utilization : Plant used as an incense in the preparation of 'dhoop', agarbatties, etc.

References : Ambasta (1986), Chopra *et al.* (1980), Kaul (1997).

373.*Morinda angustifolia* Roxb.

Syn : M. squarrosa Ham.

Family : Rubiaceae.

Distribution : Tropical Himalaya; Nepal eastward to Assam, the Khasia and Naga Hills.

Part/s Used : Flowers.

Utilization : Flowers fragrant.

References : Borthakur (1981), Watt (1972).

374.*Moringa oleifera* Lamk.

Syn : Hyperanthera decandra Willd.; *H. moringa* Vahl.; *Guilandina moringa* Linn.; *M. polygona* DC.; *M. pterygosperma* Gaertn.; *M. zeylanica* Pers.

Family : Moringaceae.

English, Sanskrit and Hindi Names: **Eng. :** Drumstick Tree, Horseradish Tree; **Sans. :** Akshiba, Bahumula, Dvisigru, Gandhaka, Janapriya, Shobhanjana; **Hindi :** Mungna, Sainjna, Shajna, Sondna.

Distribution : Throughout India; Jammu and Kashmir to Bengal, Kutch, Gujarat. Cultivated elsewhere.

Part/s Used : Roots. Seeds.

Utilization : Roots used as a condiment like true Horse-radish (*Cochlearia armoracia* Linn.) and oil from seeds, 'Ben' used in cosmetics.

Active Principles : The active principles are maringin and terigospermin. moringyne, 4-(α-L-rhamnosyloxy) benzyl isothio-cyanate and 4-(α-L-rhamnosyloxy) phenyl acetonitrile (seeds).

References : Ambasta (1986), Anonymous (1999), Bennett (1987), Kirtikar and Basu (1984), Rastogi and Mehrotra (1991, 1993, 1995), Sahni (1998), Sinha and Sinha (2001), Watt (1972).

375.*Morus nigra* Linn.

Family : Moraceae.

English and Hindi Names : **Eng. :** Black Mulberry Tree, Common Mulberry, Mulberry; **Hindi :** Shahtut.

Distribution : Cultivated.

Part/s Used : Fruits.

Utilization : Ripe fruits well flavoured and used in the preparation of a wine.

References : Ambasta (1984), Kirtikar and Basu (1984).

376.*Muntingia calabura* Linn.

Family : Tiliaceae.

English Name : **Eng. :** Japanese Cherry.

Distribution : Cultivated in Tropical regions.

Part/s Used : Leaves.

Utilization : Infusion of leaves used as tea.

Active Principles : Leaves rich in kaempferol, quercetin and their 3-O-galactosides, caffeic and ellagic acids chrysin, 2',4'-dihydroxychalcone, galangin 3,7-dimethyl ether, 5,7-dihydroxy-8-methoxyflavonol, tiliroside and buddlenoid A .

References : Ambasta (1986), Rastogi and Mehrotra (1993).

377. *Murraya koenigii* (Linn.) Spreng. (Pl. 8E)

Syn : Bergera koenigii Linn.

Family : Rutaceae.

English, Sanskrit and Hindi Names : **Eng. :** Curry Leaf Tree; **Sans. :** Alakavhaya, Chhardighna, Girinimba, Krishnapatra, Mahanimba, Nimbapatra, Surabhininiba; **Hindi :** Barsanga, Gandhela, Harri, Katnim, Kurry Patta, Mithaneem.

Distribution : Himalaya from Kumaon to Sikkim, upto 1,700m and warmer regions throughout India.

Part / s Used : Leaves.

Utilization : Leaves extensively used as a spice and condiments for flavouring curries, chutneys, soups, pickles, etc. Oil aromatic with spicy note.

Active Principles : Leaves contain girinimbine, mahanimbine, isomahanimbine, murrayacine, mahanine, koenine, koenigine, koenidine and koenimbine whereas its oil rich in β-phellandrene (24.4), α-pinene (17.5), β-caryophyllene (7.3) and terpinen-4-ol (6.1%).

References : Ambasta (1986), Bhattacharjee (1998), Chandra (1997), Chevallier (1996), Dam and Hajra (1981), Datar and Vartak (1975), Kirtikar and Basu (1984), Lal *et al.* (1996), Mehtani and Ingle (1999), Negi and Gaur (1994), Rao and Henry (1995), Rastogi and Mehrotra (1990, 1991, 1993, 1995), Saklani and Jain (1964), Watt (1972).

378. *Murraya paniculata* (Linn.) Jack

Syn : Chalcas paniculata Linn.; *M. exotica* Linn.

Family : Rutaceae.

Hindi Names : **Hindi :** Bibsar, Juti, Marchula.

Distribution : Shady ravine slopes of Garhwal Himalaya.

Part / s Used : Flowers. Bark.

Utilization : Fragrant flowers and bark universally used as a cosmetic.

Active Principles : These are murrayin, scopolin, scopoletin, glucose (flowers), coumarin-mexoticin, mexolide, sitosterol-β-D-galactoside and koenimbine (stem-bark).

References : Bennett (1987), Bhattacharjee (1998), Kirtikar and Basu (1984), Rastogi and Mehrotra (1990, 1991, 1993, 1995), Watt (1972).

379. *Myrica cerifera* Linn.

Family : Myricaceae.

English Name : **Eng. :** Candle Berry.

Part/s Used : Leaves.

Utilization : Leaves fragrant.

Reference : Bhattacharjee (1998).

380.*Myricaria germanica* Desv. ssp. ***alopecuroides***

Family : Tamaricaceae.

Distribution : Temperate and Alpine Himalaya; Sikkim to Kumaon: 3,350-4,650m.

Part/s Used : Leaves. Flowers.

Utilization : Powdered leaves and flowers used as incense.

References : Chopra *et al.* (1956), Sood *et al.* (2001).

381.*Myristica fragrans* Houtt. (Pl. 8F)

Syn : M. aromatica Lamk.; *M. officinalis* Linn.f.; *M. moschata* Thunb.

Family : Myristicaceae.

English, Sanskrit and Hindi Names : **Eng. :** False Aril, Fragrant Nut-Tree, Mace-Tree, Nutmeg Tree, True Nutmeg; **Sans. :** Jaiphala, Jatiphala, Kosha; **Hindi :** Jaiphal (Fruit Kernel), Japatri, Jayapatri (Aril), Jotri.

Distribution : Himalaya. Grown in southern states of India particularly Tamil Nadu.

Part/s Used : Fruits.

Utilization : Fruits used whole or ground as culinary spice in bakery, confectionery items, flavouring desserts, meat seasonings, sausage, fillings, sauces, pickles, soups and curries. Nutmeg (seeds) and Mace oil (dried arillode of the ripe seeds) used for flavouring food products, liquors and also cosmetics. Nutmeg butter or fat obtained from butter used in spicy perfumes and in soaps.

Active Principles : Flavour is due to myristicin and elimicin. Seed oil contains sabinene 15-30%, β-pinene 10-15%, terpinen-4-ol 6-10%, myristicin 5-8% and elimicin about 2% whereas arils rich in dehydrodiisoeugenol.

References : Ambasta (1986), Anonymous (1999), Chevallier (1996), Chopra *et al.* (1980), Greenish (1999), Harborne and Baxter (2001), Kirtikar and Basu (1984), Mehtani and Ingle (1999), Rastogi and Mehrotra (1993), Sinha and Sinha (2001), Watt (1972).

382.*Myroxylon balsamum* (L.) Harms. var. ***balsamum***

Syn : M. toluiferum H.B. and K.

Family : Papilionaceae.

English Name : **Eng. :** Tolu Balsam Tree.

Distribution : Grown in gardens as ornamental.

Utilization : It yields oil of Tolu Balsam, used as a fragrance component in soaps and in cosmetics and perfumery.

Active Principles : Tolu balsam contains 7.5% of an oily liquid (consisting of benzyl benzoate with a little benzyl cinnamate), traces of vanillin, cinnamic acid and resin.

References : Ambasta (1986), Bhattacharjee (1998), Greenish (1999), Rastogi and Mehrotra (1990,1995).

383.*Myroxylon pereirae* Klotzsch

Family : Papilionaceae.

English Name : **Eng. :** Peru Balsam Tree.

Part/s Used : Fruits. Leaves.

Utilization : Fruits and leaves contain an essential oil having a fragrant balsamic odour. Balsam oil much valued in perfumes, soaps and cosmetics.

Active Principles : Balsam of Peru consists essentially of an oily fluid portion a dark resin. The fluid portion, cinnamicin (56-66%) consists of benzyl benzoate and benzyl cinamate in the ratio of 3:2 whereas resinous portion (28%) comprises peruviol peru-resinotannol, combined with cinnamic and a little benzoic acid.

References : Ambasta (1986), Chevallier (1996), Greenish (1999).

384.*Myrtus communis* Linn.

Family : Myrtaceae.

English and Hindi Names : **Eng. :** The Common Myrtle; **Hindi :** Murad, Vilayati Mehndi.

Distribution : N.W. Himalaya. Often grown in gardens throughout India.

Part/s Used : Leaves. Twigs. Flowers. Berries (Fruits).

Utilization : Aromatic leaves and berries used for flavouring wines and food. Leaves, twigs and flowers yield an essential oil, Myrtle oil, used for flavouring meat sauces, seasonings, scenting soaps; and in perfumery under the name of Eau d' Ange.

Active Principles : These are myricetin-3-galactoside, myricetin-rhamnoside, quercetin-3-galactoside, 3-glucoside, esculin, esculatin, caffeic acid, kaempferol, myricetin, its 3' glucoside, 3-rhamnoglucoside and 3,3'-digalactoside, 2-acylphliroglucinols-myrtucommulones C and D (leaves), myricetin, kaempferol, ellagic acid and gallic acid (leaves and flowers), delphinidin, petunidin, malvidin, peonidin, cyanidin-3-mono- and 3,3'-digalactoside (berries), lauric, myristic, palmitic, stearic, oleic, linoleic and linolenic acids, α-pinene, myrcene, phellandrene, limonene, terpinene, p-cymene, linalool, linalyl acetate, β-caryophylene, α-terpineol and methyl eugenol (essential oil).

References : Ambasta (1986), Chevallier (1996), Harborne and Baxter (2001), Mehtani and Ingle (1999), Rastogi and Mehrotra (1991, 1993, 1995), Sinha and Sinha (2001), Watt (1972).

385.*Narcissus jonquilla* Linn.

Family: Liliaceae.

English Name : **Eng. :** Jonquil.

Distribution : Ornamental, grown in gardens.

Part/s Used : Flowers.

Utilization : Yield an essential oil, Jasmone valued for use in high grade perfumes.

Active Principles : Goleptine and golceptine isolated from flowers.

References : Ambasta (1986), Rastogi and Mehrotra (1990).

386.*Narcissus tazetta* Linn.

Family: Amaryllidaceae.

English Name : **Eng. :** Polyanthus Narcissus.

Distribution : W. Asia. Abundant on some of the hill sides below Shimla.

Part/s Used : Flowers.

Utilization : Flowers used in high grade perfumery.

Active Principles : These are isorhamnetinm, tazettin (flowers), α-pinene (0.08), β-pinene (0.02), car-3-ene (0.02), sabinene (0.2), lmonene (0.15), γ-terpinene (0.16), cineole (3.0), p-cymol (0.17), terpinolene (0.15), α-terpineol (0.95), linalool (16.02), nerol (0.74), geraniol (1.25), cinnamyl alcohol (5.78), methyl anthranilate (15.1) and benzyl acetate (12.25%) (absolute oil from flowers).

References : Ambasta (1986), Rastogi and Mehrotra (1991), Watt (1972).

387.*Nardostachys grandiflora* DC.

Syn.: N. jatamansi (D.Don) DC.; *Patrinia jatamansi* Don; *Valeriana jatamansi* Wall.

Family : Valerianaceae.

English, Sanskrit and Hindi Names : **Eng. :** Spikenard; **Sans. :** Akashamansi, Bhutajata, Gauri, Jatala, Mansi, Nalada; **Hindi :** Balchir, Jatamansi, Kanuchara.

Distribution : Alpine Himalaya: 3,650-5,000m.

Part/s Used : Whole Plant.

Utilization : Plant used as an incense. Also yields an essential oil.

Active Principles : Nardostachnol (69.4), 9-dehydroaristolene (0.2), 1-(10)-dehydroaristolene, 2-β-maaleine (2.9) and 1,2,9,10-tetrahydro-aristolene (2.9%) isolated from essential oil.

References : Bhattacharjee (1998), Dam & Hajra (1981), Kirtikar and Basu (1984), Rastogi and Mehrotra (1993), Saklani and Jain (1994), Watt (1972).

388.*Nasturtium officinale* R.Br.

Syn.: Cardamine fontana Lam.; *N. fontanum* Aschers.; *Roripa nasturtium – aquaticum* (L.) Hayek.

Family: Asteraceae.

English Names : **Eng. :** Shamrock, Stertion, Watercress.

Distribution : Cultivated in many other parts of India. Semi-wild-state near hill-stations on the Himalaya.

Part/s Used : Leaves.

Utilization : Leaves used for flavouring salads.

Active Principles : Flavour of leaves attributed to glucosinolates.

References : Bhattacharjee (1998), Chevallier (1996), Harborne and Baxter (2001), Kirtikar and Basu (1984), Raju (2000), Rastogi and Mehrotra (1991), Watt (1972).

389.*Nelumbo nucifera* Gaertn.

Syn.: Cyamus nelumbo Smith.; *Nelumbium nelumbo* Druce; *N. speciosum* Willd.; *N. indica* Poir.; *N. nuciferum* Gaertn.; *Nymphaea nelumbo* Linn.; *N. asiaticum* Rich.

Family: Nymphaeaceae.

English, Sanskrit and Hindi Names : **Eng.:** Chinese water lily, Egyptian Bean, Indian Lotus, Pythagorean Bean, Sacred Lotus; **Sans.:** Ambuja, Bisakusuma, Drishopadma, Harivetra, Kamala, Padma, Pankaja; **Hindi:** Ambuj, Kamal, Kanwal, Lalkamal, Padam.

Distribution : Distributed throughout the warmer parts of India and extending as far as North of Kashmir.

Part/s Used : Flowers.

Utilization : Flowers fragrant and valued in perfumery.

Active Principles : Nelumbine, 1,4-dimethoxybenzene, 1,8-cineole, terpinen-4-ol and linalool isolated from flowers.

References : Ambasta (1986), Anonymous (1999), Bennett (1987), Bhattacharjee (1998), Kaul (1997), Kirtikar and Basu (1984), Raju (2000), Rastogi and Mehrotra (1993,1995), Singh and Pandey (1998), Watt (1972).

390.*Neolitsea cassia* (Linn.) Kostermans

Syn.: Laurus cassia Linn.; *L.involucrata* Vahl.; *L. zeylanica* Herm.; *Litsea foliosa* Nees.; *L. furfuracea* Nees.; *L. oblónga* Nees.; *L. scrobiculata* Meissn.; *L. striolata* Blume; *L. trinervia* Juss.; *L. zeylanica* Nees; *N. involucrata* (Lam.) Alston; *N. zeylanica* (Nees.) Marrill.; *Tetradenia ceylanica* Nees.; *T. foliosa* Nees.; *T. furfuracea* Nees.; *Tetranthera foliosa* Wall.; *T. furfuracea* Wall.; *T. pulcherrima* (in part) Wall.

Family: Lauraceae.

Distribution : Khasia mountains. W. Coast from the Konkan southwards. Nilgiris.

Part/s Used : Leaves.

Utilization : Essential oil from leaves, called as Bellary leaf oil has odour resembling that of unripe mangoes and used for the same purposes.

Active Principles : Linalool (54.91) and β-caryophyllene (17.42%) isolated from leaf essential oil.

References : Ambasta (1986), Bennett (1987), Rastogi and Mehrotra (1991), Watt (1972).

391.*Nepeta cataria* Linn.

Family: Lamiaceae.

English Names : **Eng. :** Catnip, Catmint.

Distribution : Temperate W. Himalaya.

Part/s Used : Leaves. Shoots. Flowering Tops.

Utilization : Leaves, shoots and flowering tops employed for flavouring soups, stews and sauces. Shoots yield about 0.2 – 0.3% volatile oil.

Active Principles : Leaves rich in nepetoglucosylester. Volatile oil from shoots comprehends α – and β – nepetalactone, citronellol, geraniol, nepatolactone, epinepatolactone and caryophyllene.

References : Ambasta (1986), Bhattacharjee (1998), Chevallier (1996), Mehtani and Ingle (1999), Rastogi and Mehrotra (1990, 1993, 1995).

392.*Nepeta discolor* Royle *ex* Benth.

Family: Lamiaceae.

Utilization : Used for making incense.

Reference : Singh *et al.* (1996).

393.*Nepeta longibracteata* Benth.

Family: Lamiaceae.

Part/s Used : Leaves

Utilization : Leaves employed to impart flavour to soups.

Reference : Srivastava and Shyama (1963).

394. ***Nerium indicum*** Mill.

Syn.: N. latifolium Mill.; *N. odorum* Lamk.; *N. odorum* Soland.

Family: Apocynaceae.

English, Sanskrit and Hindi Names : **Eng.** : Indian Oleander, Sweet scented oleander; **Sans.** : Asvamaraka, Chandata, Hayamaraka, Karavira, Pratihasa; **Hindi** : Kanel, Kaner, Karber, Karuvira.

Distribution : Lower slopes of the W. Himalaya. Hilly parts of C. and S. India.

Part/s Used : Flowers.

Utilization : Flowers fragrant.

Active Principles : Flowers contain mainly oleandrin, odoroside A, kaempferol-3-glucoside, ursolic and oleanolic acids and β-sitosterol.

References : Ambasta (1986), Anonymous (1999), Bennett (1987), Bhattacharjee (1998), Kirtikar and Basu (1986), Rastogi and Mehrotra (1991), Watt (1972).

395. ***Nicotiana alata*** Link and Otto

Syn.: N. affinis Hort.; *N. persica* Lindl.

Family: Solanaceae.

Distribution : Grown in Indian Gardens.

Part/s Used : Flowers.

Utilization : Valued for its pretty fragrant flowers.

Active Principles : Nicotine, 3-oxo- α-ionol-β-glucoside and 5,6-epoxy-5,6-dihydro-3-hydroxy-β-ionol-β-glucoside isolated from flowers.

References : Ambasta (1986), Rastogi and Mehrotra (1993).

396. ***Nicotiana tabacum*** Linn.

Family: Solanaceae.

English, Sanskrit and Hindi Names : **Eng.** : Tobacco, Virginian Tobacco; **Sans.** : Dhumrapatrika, Kalanja, Krimighni, Ksharapatra, Tamakhu; **Hindi** : Bujjerbhang, Tumak, Tamaku, Tambaka, Tambaku.

Distribution : Cultivated in all tropical countries.

Part/s Used : Leaves.

Utilization : Tobacco absolute extracted from the leaves used for imparting tobacco aroma to materials used in packing tobacco products and lower grade tobaccos.

Active Principles : Tobacco flavour constituents include -(1-methyl4-isopropyl-7,8-dioxabicyclo [3.2.1.] octan-6-yl) methyl ketone (VII), 1-

(1-methyl-4-isopropyl-7,8-dioxabicyclo [3.2.1.] octan-6-yl) ethanol (VIII), 2-(1-methyl-4-isopropyl-7,8-dioxabicyclo [3.2.1.] octan-6-yl) propane-2-ol (IX), 3,3,5-trimethyl-8-isopropyl-4,9-dioxabicyclo [3.3.1.] nonan-2-ol (X), 5-isopropyl-6,7-epoxy-8-hydroxy-8-methyl-nonan-2-one (XI), 3,4-epoxy-5-isopropyl nonan-2,8-dione (XII) and trans-5-isopropyl-8-hydroxy-8-methylnon-6-en-2-one (XIII). Leaves contain scopoletin, propional, isobutyral and diethyl ketone.

References : Ambasta (1986), Bhattacharjee (1998), Chevallier (1996), Kirtikar and Basu (1984), Rastogi and Mehrotra (1991, 1993, 1995), Watt (1972).

397.*Nigella damascena* Linn.

Family: Ranunculaceae.

English Name : **Eng. :** Love in a mist.

Distribution : Grown in Indian gardens.

Part/s Used : Seeds.

Utilization : Aroma of seeds akin to strawberries and it yields a volatile oil, called Nigella Oil, used in perfumery.

Active Principles : It contains mainly damascenine, damascinine, (+) elemene and (-) α-selinene (seeds), linoleic (45.75), oleic (33.28), palmitic (7.84), eicosadienoic (4.24) and capric (3.28%) acids.

References : Ambasta (1986), Rastogi and Mehrotra (1991).

398.*Nigella sativa* Linn. (Pl. 9A)

Syn.: N. indica Roxb.

Family : Ranunculaceae.

English, Sanskrit and Hindi Names : **Eng. :** Black cumin, Nutmeg Flower, Small Fennel; **Sans. :** Bashpika, Kalajaji,karava, Karavi, Prithvi, Sushavi, Susavi; **Hindi :** Kalajira, Kalonji, Mugrela.

Distribution : Punjab. H.P. Bihar. Assam (an occasional weed of cultivation).

Part/s Used : Seeds.

Utilization : Seeds used in the flavouring of curries and other dishes.

Active Principles : These are nigellimine, esters of dehydrostearic and linoleic acids with sesqui-and higher terpenoids, aliphatic alcohols, (+) limonene, p-cymene, citronellyl acetate, carvone and (+) citronellol, cholesterol, campesterol, stigmasterol, β-sitosterol and α–spinasterol (seed oil).

References : Ambasta (1986), Anonymous (1999), Chevallier (1996) Kirtikar and Basu (1984), Mehtani and Ingle (1999), Rastogi and Mehrotra (1990, 1991, 1993, 1995), Watt (1972).

399.*Nigella reticulatus* (Stapf) Bremek.

Syn.: Strobilanthes reticulatus Stapf

Family: Ranunculaceae.

Part / s Used : Fruiting-heads.

Utilization : Fruiting-heads yield an essential oil with musty flavour.

Reference : Ambasta (1986).

400.*Nyctanthes arbor – tristis* Linn. (Pl. 9B)

Syn.: Scabrita scabra Linn.; *S. triflora* Linn.

Family: Oleaceae.

English, Sanskrit and Hindi Names : **Eng. :** Coral Jasmine. Indian Mourner, Night Flowering Jasmine, Sorrowful Tree; **Sans. :** Atyuha, Kharapatraka, Parijata, Ragapushpi, Sephalika; **Hindi :** Binari, Har, Harsinghar, Karasli, Kuri, Seoli.

Distribution : Cultivated in many parts of India.

Part / s Used : Flowers.

Utilization : Fragrant flowers jasmine scented and employed in perfumes.

Active Principles : Flowers contain mainly β-digentiobioside ester of α-crocetin crocin-3-β-monogentiobioside ester of α-crocetin and D-mannitol.

References : Ambasta (1986), Kirtikar and Basu (1984), Raju (2000), Rastogi and Mehrotra (1991), Sahni (1998), Singh and Pandey (1998), Watt (1972).

401.*Ocimum americanum* Linn.

Syn.: O. africanum Lour.; *O. album* Roxb. (not of Linn.); *O. canum* Sims

Family: Lamiaceae.

English, Sanskrit and Hindi Names : **Eng. :** Hoary Basil; **Sans. :** Ajaka, Arjaka, Gambhira, Jambira, Kuthera, Mukharjaka, Ugragandha; **Hindi :** Kala tulsi, Mamri.

Distribution : Distributed in Jammu and Kashmir, Punjab, H.P., Uttar Pradesh and Delhi. Found in plains and lower hills of India from Assam, Bengal, Bihar and C. India to the S. Deccan.

Part / s Used : Leaves.

Utilization : Fragrant leaves employed to impart flavour chutneys, vegetables, salads, pulses, tea, sauces and soups. Also, yields a volatile oil used in soaps and cosmetics.

Active Principles : These are nevadensin, salvigenin, ursolic and oleanolic acids (leaves), camphor (60.0), limonene (13.0%), bergamotene, β-caryophyllene, citral, methyl chavicol, 1-octen-3-ol, 3-octanol and eugenol (leaf oil).

References : Ambasta (1986), Bennett (1987), Bhattacharjee (1998), Kirtikar and Basu (1984), Lal *et al.* (1996), Rastogi and Mehrotra (1993, 1995), Watt (1972).

402. *Ocimum basilicum* Linn.

Family: Lamiaceae.

English, Sanskrit and Hindi Names : **Eng. :** Basil, Common Basil, Common Sweet Basil, Sweet Basil; **Sans. :** Ajagandhika, Barba, Munjariki, Surasa, Varvara; **Hindi :** Babui Tulsi, Babul, Bahari, Barbar, Gulal Tulsi, Kali Tulsi, Marua.

Distribution : Cultivated throughout the greater part of Tropical India.

Part/s Used : Leaves.

Utilization : Extensively used for flavouring confectionery, baked goods, sauces, pickles, fancy vinegars, spiced meats, sausages, beverages, ketchups, tomato pastes and all major food products. Essential oil is of two types: (i) Sweet basil oil and (ii) Exotic basil oil. Former is used in perfumery, cosmetics, detergents, dental creams and mouth washes, while the latter finds use in high-class fragrances, soaps, detergents and dental products.

Active Principles : Chief constitutents of the essential oil are methyl cinnamate, linalool or methyl chavicol, ocimene, linalyl acetate, eugenol and trans – anethole, methyl cinnamate, cineole and other terpenes. Sweet basil has 15-50% linalool and 15-25% methyl chavicol whereas methyl chavicol content of exotic basil oil is greater than 80%.

References : Ambasta (1986), Bhattacharjee (1998), Chevallier (1996), Jain and Sharma (2000), Harborne and Baxter (2001), Kirtikar and Basu (1984), Mehtani and Ingle (1999), Rastogi and Mehrotra (1990, 1991, 1993, 1995), Watt (1972).

403. *Ocimum gratissimum* Linn.

Family : Lamiaceae.

English, Sanskrit and Hindi Names : **Eng. :** Large Basil, Lemon Basil, Shrubby Basil; **Sans. :** Ajaka, Nidralu, Ramatulasi, Sugandhi, Vriddhutulsi; **Hindi :** Bantulsi, Ram Tulsi.

Part/s Used : Whole Herb.

Utilization : Herb more strongly scented and used in aromatic bath for rheumatism and paralysis. Also, yields an essential oil.

Active Principles : These are ocimol, gratissimin (leaves), eugenol (63.5), ocimene (14.8), (-)-cadinene (4.0) and (-) perillyl alcohol (3.1%) (oil).

References : Ambasta (1986), Kirtikar and Basu (1984), Rastogi and Mehrotra (1990,1993), Watt (1972).

404. *Ocimum kilimandscharicum* Guerke

Family: Lamiaceae.

English, Sanskrit and Hindi Names : **Eng. :** Camphor Basil; **Sans. :** Karpur Tulsi; **Hindi :** Kapur Tulsi.

Distribution : Introduced and Cultivated in W. Bengal, Jammu and Kashmir, Dehradun and Lucknow.

Part/s Used : Leaves. Flowers.

Utilization : Leaves contain maximum amount of camphor and oil, followed by flowers.

Active Principles : Essential oil contains mainly camphor with content varies from 61- 80.5%.

References : Ambasta (1986), Bhattacharjee (1998), Rastogi and Mehrotra (1995).

405. *Ocimum sanctum* Linn. (Pl. 9C)

Family: Lamiaceae.

English, Sanskrit and Hindi Names : **Eng. :** Holy Basil, Monk's Basil, Rough Basil, Sacred Basil; **Sans. :** Ajaka, Bana Tulsi, Brinda, Divya, Gauri, Haripriya, Laxmi, Manjari, Parnasa, Patrapuspha, Suvasa Tulasi, Vrinda; **Hindi :** Baranda, Kala Tulsi, Tulsi, Varanda.

Distribution : Himalaya upto 2,000m.

Part/s Used : Leaves.

Utilization : Leaf oil extensively used for flavouring food products, confectionery goods, condiments and toiletry products (mouth washes, dental creams).

Active Principles : Essential oil contains carvacrol, caryophyllene, nerol and camphene, etc.

References : Ambasta (1986), Anonymous (1999), Bhattacharjee (1998), Chevallier (1996), Kirtikar and Basu (1984), Raju (2000), Rastogi and Mehrotra (1991, 1993, 1995), Sinha and Sinha (2001), Watt (1972).

406. *Ocimum viride* Willd.

Family: Lamiaceae.

English Name: **Eng. :** Fever plant of Sierre Leone.

Distribution : Introduced into India.

Part/s Used : Leaves.

Utilization : Plant highly scented and used as a flavouring agent.

Active Principles : α-pinene (1.40), limonene (1.30), 1,8-cineole (3.40), p-cymene (6.70), linalool (3.4), thymol (58.15), linalyl acetate (1.8), β-caryophyllene (3.98) and β-caryophyllene oxide (2.58%) isolated from oil.

References : Ambasta (1986), Mehtani and Ingle (1999), Rastogi and Mehrotra (1993).

407. *Oenanthe javanica* (Blume) DC.

Syn. : *O. benghalensis* (DC.) Benth. and Hook.f.; *O. javanica* ssp. *linearis* (Wall. *ex* DC.) Murata; O.j. ssp. *stolonifera* (Wall. *ex* DC.) Murata; *O. linearis* Wall. *ex* DC.; *O. stolonifera* (Roxb.) Wall. *ex* DC.; *O. thomsonii* Cl.; *Sium javanicum* Bl.

Family: Apiaceae.

Distribution : Kashmir to Assam ascending to an altitude of 1800m. Plains of Bengal.

Part/s Used : Fruits. Leaves. Flowering tops.

Utilization : Eaten as a flavouring. Fruits yield an essential oil. Leaves used as condiments to flavour curries.

Active Principles : Dillapiole, α-pinene, β-phellandrene, p-cymene, 1,8-cineole, β-caryophyllene, caryophyllene oxide (essential oil), palmitic acid, aspartic acid, stigmasterol, glucosamine and galactosamine from (aerial parts), persicarin, quercetin and rutin from petals.

References : Ambasta (1986), Arora (1981), Bennett (1987), Chopra *et al.* (1980), Mehtani and Ingle (1999), Mittre (1991), Rastogi and Mehrotra (1990, 1991, 1993).

408. *Olax scandens* Roxb.

Syn.: *O. bador* Ham., Roxburghia Koenig.

Family: Olacaceae.

Hindi Name: **Hindi :** Dheniani.

Distribution : Tropical forests of the W. Himalaya from Kumaon to Assam and extending into C. and S. India.

Part/s Used : Flowers. Fruits.

Utilization : Flowers fragrant.

References : Ambasta (1986), Kirtikar and Basu (1984), Negi and Gaur (1994), Watt (1972).

409. *Oncosperma tigillarium* (Jack) Ridley

Syn. : *O. filamentosum* Blume

Family: Arecaceae.

Part/s Used : Bud.

Utilization : Buds used for flavouring.

Reference : Ambasta (1986).

410. *Origanum vulgare* Linn.

Family: Lamiaceae.

English and Hindi Names: **Eng.** : Common or wild marjoram; **Hindi** : Sathra.

Distribution : Temperate Himalaya from Kashmir to Sikkim.

Part/s Used : Leaves. Flowers. Aerial parts.

Utilization : Aerial parts including leaves and flowers used as spices and condiments for flavouring. Also, yield a volatile oil used in perfumery, soap industry, and to flavourable and liquors.

Active Principles : Herb yields 0.15–0.40% volatile oil comprising carvacrol, thymol, beta-bisabolene, caryophyllene, linalool and borneol whereas 1,8-cineole, limonene, terpineol, thymol, α-thujene, camphene, β-pinene, myrcene, α-phellandrene, α-terpinene, ocimene, p-cymene, carvacrol and γ-terpinene isolated from essential oil.

References : Ambasta (1986), Bhattacharjee (1998), Chandra (1997), Chevallier (1996), Kaul (1997), Kirtikar and Basu (1984), Mehtani and Ingle (1999), Rastogi and Mehrotra (1990, 1991, 1993, 1995), Watt (1972).

411. *Oryza meyeriana* (Zoll. and Mor. *ex* Steud.) Baill.

Syn. : *O. granulata* Nees and Arn. *ex* Steud

Family: Poaceae.

Distribution : Found on dry soils upto 1,000m.

Part/s Used : Grains.

Utilization : Grains possess good flavour.

References : Ambasta (1986), Watt (1972).

412. *Osmanthus fragrans* Lour.

Syn. : *Olea acuminate* Wall.; *O. fragrans* Thunb.

Family: Oleaceae.

Distribution : Himalaya: 1,000 – 2,900m. Hills of Assam and Uttar Pradesh.

Part/s Used : Flowers. Leaves.

Utilization : Both flowers and leaves employed to impart flavour to tea. Flowers also used for flavouring confectionery, bakery products and wines, and essential oil also extracted from them.

Active Principles : Chief ingredients of essential oil from flowers are osmane, trans-linalool oxide, cis-linalool oxide, α-and β- ionones, nerol, linalool, geraniol, γ-decalactone, 1-hexanol, megastigma-6,8-dien-2,5-oxide, nonanol, dihydro-β-ionone, ethyl palmitate, 1,2-dimethylbenzene, 6-methyl-5-hepten-2-one, 3-hexenyl formate, α-terpineol, 3,4-dimethyl heptane, trans-6-ethenyl tetrahydro-2,2,6-trimethylpyran-3-ol, cis-6-ethenyl tetrahydro-2,2,60 trimethylpyran-

3-ol, benzylethanol, 3,7,11-trimethyl-1,6,10-dodecatrien-3-ol, m-ethylphenol, 9,12-octadecadienol, methyl 3-hydroxy benzoate and methyl 11, 14, 17-eicosatrienoate.

References : Ambasta (1986), Chandra (1997), Chopra *et al.* (1980), Mehtani and Ingle (1999), Rastogi and Mehrotra (1991,1993), Watt (1972).

413. ***Osmorhiza aristata*** (Thunb.) Makino and Yabe var. ***laxa*** (Royle) Constance and Shan

Syn. : *O. claytonii* C.B. Clarke in part; *O. laxa* Royle

Family: Apiaceae.

Part/s Used : Roots.

Utilization : Aromatic roots used for flavouring.

Reference : Ambasta (1986).

414. ***Ottelia alismoides*** Pres.

Family: Hydrocharitaceae.

Part/s Used : Leaves.

Utilization : Leaves sweetly fragrant.

Reference : Ambasta (1986).

415. ***Oxalis acetosella*** Linn.

Family: Oxalidaceae.

English Names : **Eng.:** Allolida, Bird's Clover, Common Wood-Sorrel, Cuckoo's Clover, French Sorrel, Sleeping Beauty, Sour Clover.

Distribution : Temperate Himalaya-Kashmir to Sikkim: 2,750-4,000m.

Part/s Used : Leaves.

Utilization : Leaves have refreshing flavour and used as a substitute for sorrel (*Rumex acetosa* Linn.) in salads.

Active Principles : Leaves rich in vitamin-C, potassium oxalate and -2"-O-(β-D-glucopyranosyl) isovitexin.

References : Ambasta (1986), Bhattacharjee (1998), Kirtikar and Basu (1984), Rastogi and Mehrotra (1991), Watt (1972).

416. ***Oxalis maritiana*** Zucc.

Syn.: *O. corymbosa* DC.

Family: Oxalidaceae.

Distribution : Naturalised in moist and shady places in Temperate parts and Palni hills in S. India.

Part/s Used : Tubers.

Utilization : Tubers possess a pleasant flavour.

References : Bennett (1987), Chandra (1997).

417. ***Paeonia emodi*** Wall *ex* Royle

Syn : P. officinalis H.f. and T.

Family : Ranunculaceae.

English and Hindi Names: **Eng. :** Himalayan Paeony, Paeony Rose; **Hindi :** Ud-salap.

Distribution: W. Temperate Himalaya - Kashmir to Kumaon: 1,700-3,400m.

Part/s Used : Roots.

Utilization : Roots contain an essential oil, used in perfumery.

Active Principles : Root oil contains C14-33n alkanes, β-amyrin, butyrospermol, cycloartenol, lupeol, 24-methylenecycloartanol, cholesterol, campesterol and sitosterol (unsaponifiable lipid), octanoic, decanoic, lauric, myristic, myristoleic, palmitic, palmitoleic, stearic, oleic and linoleic acids (saponifiable lipid).

References : Kirtikar and Basu (1984), Raju (2000), Rastogi and Mehrotra (1993), Watt (1972).

418. ***Palmaria palmata*** (Linn.) O.Kuntze

Family : Rhodomeniaceae.

English Name : **Eng. :** Dulse.

Part/s Used : Algal Thallus.

Utilization : Used as a condiment and for flavouring alcoholic beverages.

Reference : Ambasta (1986).

419. ***Pandanus odoratissimus*** Linn.f. (Pl. 10A)

Syn : P. amaryllifolius Roxb.; *P. fascicularis* Lamk.; *P. laevis* Kunth; *P. latifolius* Hassk.; *P. tectorius* Soland. *ex* Parkinson; *P. variegatus* Miq.

Family : Pandanaceae.

English, Sanskrit and Hindi Names: **Eng. :** Screw-pine, Umbrella tree; **Sans. :** Chamarapushpa, Dirghapatra, Ketaki; **Hindi :** Gagandhul, Keora, Ketgi, Ketki, Keura, Kewda.

Distribution : Andaman Islands. Sea Coast of the Indian Peninsula.

Part/s Used : Flowers. Leaves. Roots.

Utilization : Tender floral leaves used for flavouring cream, rice (giving a flavour similar to basmati rice), 'sherbets', jellies and sometimes curries. Essence of roots commercially used for flavouring sweets and cold drinks. Male flowers used for extraction of `Kewda water' and `Kewda attar', much valued by Indian perfumers. The former is mostly used for flavouring foods, sweets, syrups and whereas latter finds use for scenting clothes, bouquets, lotions, cosmetics, soaps, hair oils, tobacco and agarbattis.

Active Principles : Characteristic aroma of spadices and Kewda oil have been attributed to β-phenyl ethyl methyl ether and 2-acetyl-1-pyrroline.

References : Agrawal (1981), Ambasta (1986), Anonymous (1999), Bennett (1987), Chopra *et al.* (1956), Kirtikar and Basu (1984), Mehtani and Ingle (1999), Sahni (1998), Watt (1972).

420. ***Pandanus utilis*** Bory

Family : Pandanaceae.

Part/s Used : Male spadices.

Utilization : Male spadices pleasantly odorous.

Reference : Ambasta (1986).

421. ***Parkia roxburghii*** G.Don.

Syn : P. javanica Merrill; *P. timoriana* (A.DC.) Merrill; *Inga timoriana* A.DC.

Family : Mimosaceae.

Distribution : Indigenous to India; mainly occurs in Assam.

Part/s Used : Pods.

Utilization : Tender pods used for flavouring foods.

References : Bhattacharjee (1998), Bennett (1987).

422. ***Parmelia abessinica*** Kremp.

Family : Parmeliaceae.

Part/s Used : Lichen Thallus.

Utilization : Used as a condiment.

Reference : Ambasta (1986).

423. ***Parmelia cirrhata*** Fr.

Family : Parmeliaceae.

Distribution : Sub-Himalayan tract of India.

Part/s Used : Lichen Thallus.

Utilization : Used as a spice and for flavouring meat, curries and vegetables.

Active Principles : Atranorin (1.5), salazinic (3.8) and usnic (0.3%) acids isolated from lichen thallus.

References : Chopra *et al.* (1980), Rastogi and Mehrotra (1993), Upreti (1996).

424. *Parmelia furfuracea* (Linn.) Ach.

Syn.: Evernia furfuracea Mann

Family : Parmeliaceae.

Part/s Used : Lichen Thallus.

Utilization : Oleoresin extracted from the lichen, used in perfumery.

Reference : Ambasta (1986).

425. *Parmelia reticulata*

Family : Parmeliaceae.

Part/s Used : Lichen Thallus.

Utilization : Used as a spice.

Reference : Upreti (1996).

426. *Parmelia tinctorium* Despr.

Family : Parmeliaceae.

Hindi Name : **Hindi :** Awasthi.

Distribution : Indian plains and hills.

Part/s Used : Lichen Thallus.

Utilization : Used as spice and flavouring agent.

Active Principles : Lichen contains stranorin and lecanoric acid isolated from the lichen.

References : Chopra *et al.* (1980), Rastogi and Mehrotra (1993), Upreti (1996).

427. *Parmotrema abessinicum*

Family: Parmeliaceae.

Distribution : Tamil Nadu.

Part/s Used : Lichen Thallus.

Utilization : Used as condiments.

Active Principles : Chief constituents are atranorin, loxodin, norlobaridone and protolichesterinic acid.

Reference : Mudgal and Hajra (1997).

428. *Parmotrema reticulatum* (Taylor) M. Choisy

Family : Parmeliaceae.

Part/s Used : Lichen Thallus.

Utilization : Used as spices and for seasoning non-vegetarian preparations, pulses and other vegetables.

Reference : Mudgal and Hajra (1997).

429. *Parmotrema tinctorium* (Despr. *ex* Nyl.) Hale.

Family : Parmeliaceae.

Part/s Used : Lichen Thallus.

Utilization : Used as spices and for seasoning non-vegetarian preparations, pulses and other vegetables.

Reference : Mudgal and Hajra (1997).

430. *Passiflora edulis* Sims.

Family : Passifloraceae.

English Names: **Eng. :** Passion fruit, Purple granadilla.

Distribution : Hilly areas of Tamil Nadu, Andhra Pradesh, W. Bengal, Assam and Nilgiris.

Part/s Used : Fruits.

Utilization : Fruit juice used for flavouring cake fillings, candy, carbonated beverages, cordials, frostings and ice-creams.

Active Principles : Flavour components of fruit juice are (6E,2'E)-6-(but-2'-enylidene)-1,5,5-trimethylcyclohex-1-ene, stereoisomers of 6-(but-2'-enylidene)-1,5,5-trimethylcyclohex-1-ene, ethyl (Z)-4,7-octadienoate, (Z)-3,5-hexadienylbutyrate, (2R*, 4S*, 4aS*, 8aS*)-4,4a-epoxy-4-4a dihydroedulan and (2R*,3S*,8aS*)-3-hydroxyedulan.

References : Ambasta (1986), Kirtikar and Basu (1984), Mehtani and Ingle (1999), Rastogi and Mehrotra (1991).

431. *Passiflora quadrangularis* Linn.

Family : Passifloraceae.

English Names: **Eng. :** Giant Granadilla, Square stalked passion flower, True Granadilla.

Distribution : Cultivated in the hotter parts of India.

Part/s Used : Fruits.

Utilization : Fruit pulp pleasantly flavoured.

References : Ambasta (1986), Watt (1972).

432. *Pavetta tomentosa* Roxb. *ex* Sm.

Syn : P. brunonis Wight; *P. canescens* Wall.; *P. indica* ssp. *tomentosa* (Roxb. *ex* Smith) Bennett; *P. indica* var. *tomentosa* (Roxb. *ex* Smith) Hook.f.; *P. mollis* Wall.; *P. velutina* Wall.; *Ixora tomentosa* DC.

Family : Rubiaceae.

Distribution : Throughout India from the W. Himalaya in Garhwal ascending to 1,350m.

Part/s Used : Flowers.

Utilization : Flowers fragrant.

References : Bennett (1987), Negi and Gaur (1994), Watt (1972).

433. *Pavonia odorata* Willd.

Syn : Hibiscus odoratus Roxb.; *Pavonia romborua* Wall.; *P. rosea* Wall.; *P. sidoides.*

Family : Malvaceae.

Sanskrit and Hindi Names: **Sans. :** Ambunamaka, Bala, Hribera, Keshya, Varida; **Hindi :** Sugandha-bala.

Distribution : N.W. India. W. Peninsula. Also, cultivated in gardens.

Part/s Used : Flowers. Roots.

Utilization : Flowers fragrant. Roots used in high grade perfumery for possessing musk-like odour. Roots also contain an essential oil.

Active Principles : Essential oil comprises isovaleric acid, n-caproic acid, α-pinene and methyl heptenone.

References : Ambasta (1986), Kirtikar and Basu (1984), Rastogi and Mehrotra (1990), Watt (1972).

434. *Pelargonium capitatum* (L.) Ait.

Family : Geraniaceae.

Distribution : Introduced into hill stations of S. India.

Utilization : Source of an essential oil used in perfumery.

Active Principles : These are α-pinene, α-phellandrene, p-cymene, γ-terpinene, β-caryophyllene, guaia-6,9-diene, germacrene D, δ-cadinene, citronellol, terpinen-4-ol, 10-epi-γ-eudesmol and citronellyl formate (essential oil).

References : Ambasta (1986), Rastogi and Mehrotra (1995).

435. *Pelargonium graveolens* L'Herit

Family : Geraniaceae.

English Name : **Eng. :** Geranium.

Distribution: Temperate, Sub-Tropical and Tropical climates. Introduced in India in the Nilgiris and Anaimalais at 900-2200m.

Part/s Used : Leaves.

Utilization : Geranium oil extracted from the leaves used for soaps, perfumes, cosmetics and for flavouring tobacco and pharmaceutical preparations, and also an excellent base for artificial rose ottos.

Active Principles : Geranium oil contains citronellol (27.54), geraniol (19.08), linalool (10.16), caryophyllene (8.07), citronellyl acetate (7.75%), vanillyl formate and vanillyl alcohol.

References : Ambasta (1986), Bhakuni (2000), Kumar and Singh (2000), Rastogi and Mehrotra (1991, 1993, 1995), Sharma and Sood (1997).

436.*Pelargonium odoratissimum* Ait.

Family : Geraniaceae.

Utilization : Essential oil used in perfumery.

Reference : Ambasta (1986).

437.*Pelargonium quercifolium* Ait.

Family : Geraniaceae.

Part/s Used : Leaves.

Utilization : Ladanum scented Geranium oil extracted from leaves.

Reference : Ambasta (1986).

438.*Perilla frutescens* (Linn.) Britton

Syn : Melissa maxima Arduin; *Mentha perilloides* Willd.; *Ocimum frutescens* Linn.; *P. macrostachya* Benth.; *P. ocimoides* Linn.

Family : Lamiaceae.

Hindi Name: **Hindi :** Bhanjira.

Distribution : Temperate and Tropical Himalaya-Kashmir to Bhutan: 350-3,350m. Khasia mountains: 1,000-2,000m.

Part/s Used : Flowering Tops. Leaves. Seeds.

Utilization : Leaves, flowering tops and seed flour used for flavouring. Yields a volatile oil, Perilla oil used for flavouring table sauces and confectionery. Seeds also yield an aromatic oil.

Active Principles : These are rosefuran, β-caryophyllene, perilla ketone, perillene, germacrene-D, α–phellandrene, α-humulene, δ-cadinene, α-terpineol, cis-α-farnesene, trans-(C10), cis-(C10)-allofarnesenes, with β-farnesene, farnesol and perillaldehyde (oil), perillyl-β-D-glucoside, monoterpene glucosides-perilloside A,B,C and D (leaves), ethyl linolenate, linolenic acid, β-sitosterol, perillene, perillaketone, egomaketone and isoegomaketone (aerial parts), oleic, linolenic and linoleic acids and sitosterol (seeds).

References : Ambasta (1986), Bennett (1987), Mehtani and Ingle (1999), Rao and Henry (1995), Rastogi and Mehrotra (1990, 1991, 1993, 1995), Watt (1972).

439. ***Perilla ocimoides*** Linn.

Family : Lamiaceae.

Part/s Used : Flowering Tops.

Utilization : Used for flavouring cordial drinks.

Reference : Gupta (1981).

440. ***Perovskia atriplicifolia*** Benth.

Family : Lamiaceae.

Distribution : Kashmir: 2,500-3,350m.

Part/s Used : Flowers.

Utilization : Flowers yield an essential oil having a camphoraceous aroma.

References : Ambasta (1986), Kirtikar and Basu (1984).

441. ***Persea americana*** Mill.

Syn : P. americana var. *drymifolia* Blake; *P. gratissima* Gaertn. f.

Family : Lauraceae.

English Names: **Eng. :** Alligator Pear, Avocado, Butter fruit.

Distribution : Grown in Poona, Bangalore and S. India including foothills of Nilgiris and Pulney hills.

Part/s Used : Bark. Leaves.

Utilization : Bark yields an essential oil with odour akin to liquorice whereas essential oil from leaves resembles Tarragon oil (*Artemisia dracunculus* Linn.). These are used in cosmetics and in toiletries.

Active Principles : Main constituents are scopoletin, apigenin, luteolin, its 7-glucoside, astragalin, quercetin-3-diglucoside (leaves), two arabinoxylans, one of which was composed of arabinose and xylose in molar ratio of 3:1 and other of xylose and arabinose in molar ratio of 19:1 (bark), methyl chavicol and α-pinene (volatile oil, leaves and bark).

References: Ambasta (1986), Bennett (1987), Chevallier (1996), Harborne and Baxter (2001), Rastogi and Mehrotra (1991, 1995).

442. ***Petiveria alliacea*** Linn.

Family : Phytolaccaceae.

Distribution : Introduced into Indian gardens.

Part/s Used : Whole Plant.

Utilization : Plant smells of garlic.

Active Principles : β-sitosterol, pinitol, allantoin, lignocerylalcohol, lignoceric acid, lignoceryl lignocerate and α-friedelinol isolated from whole plant.

References : Ambasta (1986), Rastogi and Mehrotra (1991).

443. *Petroselinum crispum* (Mill.) Airy-Shaw

Syn : P. hortense Hoffm.; *P. sativum* Hoffm.; *Apium crispum* Mill.; *A. petroselinum* L.

Family : Apiaceae.

English Name : **Eng. :** Parsley.

Distribution : Cultivated in kitchen gardens.

Part/s Used : Whole Herb.

Utilization : Dried leaves and roots used as condiments and for culinary purposes. Fresh leaves used for garnishing and seasoning, soups, stews and sauces. Yields an essential oil, Parsley herb oil, used mainly for flavouring food products (sauces, pickles, meat, additives) and various spice blends. All parts of the plant contain essential oil.

Active Principles : Main ingredients are apiole, bergapten, camphene, isopimpinellin, myristicin, α-pinene, psoralen and xanthotoxin (plant), 2-(p-tolyl) propan-2-ol (leaves), together with 4-isopropenyl-1-methyl-benzene, p-p-mentha-1,3,8-triene, β-phellandrene, imperatorin, 5-methoxypsoralen, 8-methox-ypsoralen and oxypeucedanin (leaves and roots), apiole (1.7) and myristicin (0.8%) (essential oil of fruits), myristicin (21.8), p-1,3,8-menthatriene (10.49), 1-methyl-4-isopropenyl benzene (4.52) and myrcene (3.02%) (0.06% leaf oil).

References: Ambasta (1986), Bhattacharjee (1998), Chevallier (1996), Mehtani and Ingle (1999), Rastogi and Mehrotra (1993, 1995), Watt (1972).

444. *Peucedanum dhana* Ham. *ex* Clarke

Family : Apiaceae.

Distribution : W. Himalaya upto 2,350 m. N. Bengal.

Part/s Used : Seeds.

Utilization : Seeds employed as a condiment. Also, yield an essential oil.

Active Principles : These are 5-geranoxypsoralen, imperatorin and isoimperatorin (seeds), bergapten, xanthotoxin, isoimperatorin, imperatorin, 5-geranyloxypsoralen, α-and β-pinenes, camphene, limonene, β-phellandrene, p-cymene, car-3-ene, bornyl acetate, α-terpineol, geranyl acetate, geraniol, β-caryophyllene and its oxide (seed essential oil).

References : Jain (1964), Kirtikar and Basu (1984), Rastogi and Mehrotra (1991, 1993),

445. *Peucedanum grande* C.B. Clarke

Syn : Pastinaca grandis Dals. and Gibs.

Family : Apiaceae.

Hindi Name: **Hindi :** Duku.

Distribution : W. Ghats. Hills of Deccan plateau.

Part/s Used : Fruits.

Utilization : Fruits used as condiments. Also, yield an essential oil.

Active Principles : Seeds possess columbianadin, byakangelicin and imperatorin.

References : Ambasta (1986), Chandra (1997), Kirtikar and Basu (1984), Rastogi and Mehrotra (1991), Watt (1972).

446. *Phyllanthus reticulatus* Poir.

Syn : Anisonema dubium Blume; *A. multiflora* Wight.; *A. reticulatum* A. Juss.; *A. zollingeri* Miquel; *Cicca microcarpa* Benth.; *C. reticulata* Kurz.; *Kirganelia multiflora* Thwaites.; *Rhamnus zeylanicus* Burm.; *Phyllanthus kirganelia* Herb. Ham. and Roxb.; *P. microcarpus* Muell. Arg.; *P. multiflorus* Willd.; *P. pentandrus* Herb. Roxb.; *P. prieurianus* Muell.-Arg.; *P. puberulus* Miquel.; *P. sinensis* Muell. Arg.; *P. spinescens* Wall.; *P. virosus* Wall.

Family : Euphorbiaceae.

Sanskrit and Hindi Names: **Sans. :** Bahupraja, Bahupushpa, Kamboji; **Hindi :** Buinowla, Kalemadhkaper, Makhi, Panjoli, Panjuli.

Distribution : Common throughout Tropical India.

Part/s Used : Leaves.

Utilization : Leaves used for flavouring.

References : Kirtikar and Basu (1984), Kshirsagar *et al.* (2003), Watt (1972).

447. *Phymatodes longissima* (Blume) J. Smith

Family : Polypodiaceae.

Part/s Used : Young Fronds.

Utilization : Young fronds eaten as a flavouring.

References : Ambasta (1986), Singh and Viswanathan (1996).

448. *Phymatodes scolopendria* (Burm.) Ching.

Syn : Pleopeltis phymatodes Moore.

Family : Polypodiaceae.

Part/s Used : Fronds.

Utilization : Used as a flavouring agent.

Reference : Singh and Viswanathan (1996).

449. *Physalis peruviana* Linn.

Syn : P. edulis Sims.; *P. pubescens* Don.

Family : Solanaceae.

English, Sanskrit and Hindi Names : **Eng. :** The Cape Gooseberry; **Sans. :** Tankasi; **Hindi :** Macao, Tepariyo, Tipari.

Distribution : Cultivated throughout India.

Part / s Used : Fruits.

Utilization : Fruits finely flavoured.

References : Ambasta (1986), Kirtikar and Basu (1984), Watt (1972).

450. *Phytolacca acinosa* Roxb.

Syn : P. decandra var. *β-acinosa* Moq.; *P. latbenia* (Moq.) Walter; *Pircunia kaempferi*, A. gray.; *P. latbenia* Moq.

Family : Phytolaccaceae.

English and Hindi Names: **Eng. :** Indian Poke, Sweet Belladona; **Hindi :** Matazor, Sarangum.

Distribution : Temperate Himalaya, wild or cultivated. Assam: 1,650-3,000m.

Part / s Used : Fruit.

Utilization : Fruits occasionally used as a flavouring agent.

Active Principles : Fruits contain acinosolic acid, hentriacontanol, jailigonic acid, lignoceryl palmitate, myristic acid, n-pentacosane, spergulagenic acid, β-sitosterol, ursolic acid and its galactoside.

References : Ambasta (1986), Bennett (1987), Kaul (1997), Kirtikar and Basu (1984), Mehtani and Ingle (1999), Rastogi and Mehrotra (1991), Watt (1972).

451. *Picea smithiana* Boiss.

Syn : Abies smithiana Lindl.; *A.spinulosa* Griff.; *Picea morinda* Link.

Family : Pinaceae.

English Name : **Eng. :** W. Himalayan Spruce.

Distribution : Himalaya-Kashmir to Kumaon: 2,150-3,300m.

Part / s Used : Leaves.

Utilization : Leaf oil used in bath salts, deodorants and room sprays.

Active Principles : Leaves contain agathisflavone, amentoflavone and cupressuflavone.

References : Ambasta (1986), Rastogi and Mehrotra (1995), Sahni (1998), Watt (1972).

452. *Pilea melastomoides* (Poir.) Blume

Syn : P.trinervia (Roxb.) W.; *Urtica melastomoides* Poir.

Family : Urticaceae.

Part / s Used : Leaves.

Utilization : Aromatic leaves used for seasoning.

References : Ambasta (1986), Bennett (1987).

453. *Pimenta dioica* (Linn.) Merrill.

Syn : P. officinalis Lindl.

Family : Myrtaceae.

English Names : **Eng. :** Allspice Tree, Jamaica Pepper Tree, Pimento Tree.

Distribution : Grown in the gardens of Bangalore, Bihar, Orissa, Kerela and W. Bengal.

Part/s Used : Fruits. Leaves.

Utilization : Dried unripe fruits constitute pepper or Pimento of commerce. Fruit oil used for flavouring and in perfumery. Oil from leaves, known as Pimento, used to flavour meat products and confectionery. Besides, an oleoresin extracted from pimento berries used in spice blends for the canning and meat preserving industry. Dried berries used for flavouring ketchups, soups, sauces, pickles, canned meat, fish, poultry, fast foods like hamburgers, sausages, frankfurters and mulled wines.

Active Principles : Pimento berries contain from 3-4.5% of volatile oil consisting mainly of eugenol (about 60-80%), the principle constituent of the berry oil. Besides, methyl eugenol, limonene, β-carypphyllene, 1,8-cineole, α-phellandrene, terpinolene, β-selinene, α-humulene and p-cymene are also isolated. The leaf oil contains mainly eugenol (90%), β-caryophyllene, 1,8-cineole, α-humulene and caryophyllene oxide.

References : Ambasta (1986), Chevallier (1996), Greenish (1999), Harborne and Baxter (2001), Mehtani and Ingle (1999), Rastogi and Mehrotra (1995).

454. *Pimenta racemosa* (Mill.) J.W. Moore

Syn : P. acris Kostel.

Family : Myrtaceae.

English Names: **Eng. :** Bay Rum Tree, Bay Tree.

Distribution : Introduced into Indian gardens.

Part/s Used : Leaves.

Utilization : Green leaves yield an essential oil, bay oil, used in perfumery and for flavouring table sauces.

Active Principles : Essential oil contains 1,8-cineole and eugenol.

References : Ambasta (1986), Chopra *et al.* (1980), Kirtikar and Basu (1980), Mehtani and Ingle (1999).

455. *Pimpinella anisum* Linn.

Family : Apiaceae.

English, Sanskrit and Hindi Names : **Eng. :** Aniseed, The Anise; **Sans. :** Shetapushapa; **Hindi :** Badian, Saunf, Sawonf.

Distribution : Cultivated mainly in N. India.

Part/s Used : Fruits. Seeds.

Utilization : Fruits used for flavouring beverages, confectionery and foods. Seeds yield aniseed oil, used in perfumery.

Active Principles : These are anethol, methyl chavicol and other terpene (oil from fruits) and cis-anethole, carvone, β-caryophyllene, dihydrocarvyl acetate, estragole and limonene (essential oil from seeds).

References : Ambasta (1986), Bhattacharjee (1998), Chevallier (1996), Chopra *et al.* (1960), Greenish (1999), Harborne and Baxter (2001), Mehtani and Ingle (1999), Rastogi and Mehrotra (1990, 1991, 1993, 1995), Watt (1972).

456. ***Pimpinella heyneana*** (DC) Kurz.

Family : Apiaceae.

Distribution : Konkan. Deccan. Circars.

Part/s Used : Leaves. Seeds.

Utilization : Both leaves and seeds used as condiments.

References : Kirtikar and Basu (1984), Ramachandran and Nair (1981), Saxena and Dutta (1975).

457. ***Pinus roxburghii*** Sarg.

Syn : *P. longifolia* Roxb.

Family : Pinaceae.

English, Sanskrit and Hindi Names : **Eng.** : Chir Pine, Himalayan Long Leaved Pine, Three Leaved Pine; **Sans.** : Bhadradaru, Manojna, Pitadaru, Sarala; **Hindi** : Chil, Chir, Dhupasarala, Saral, Sarala, Salla.

Distribution : All along the Himalaya: 450-2,300m.

Part/s Used : Rosin. Twigs. Needles. Cones.

Utilization : Young twigs, fresh needles and cones yield Pine needle oil, used for scenting soaps, bath preparations, room sprays and deodorants. Rosin tapped is also used principally in soap and cosmetic industries.

Active Principles : Turpentine oil contains mainly β-pinene and car-3-ene.

References : Ambasta (1986), Kirtikar and Basu (1984), Rastogi and Mehrotra (1990, 1993, 1995), Sahni (1998), Watt (1972).

458. ***Pinus smithiana*** Boiss.

Family : Pinaceae.

Part/s Used : Leaves.

Utilization : Leaf oil used in deodorants and room sprays.

Reference : Kapur (1996b).

459. *Piper betle* Linn.

Syn : Chavica betle Miq.; *C. chuvya* Miq.; *C. siriboa* Miq.; *Piper chavya* Ham.; *P. peepuloides* Wall.; *P. siriboa* Miq.

Family : Piperaceae.

English, Sanskrit and Hindi Names : **Eng. :** Betel Leaf Vine, Betel Pepper; **Sans. :** Bhakshyapatra, Divabhishta, Nagavalli, Parna, Tambula; **Hindi :** Pan, Tambuli.

Distribution : Cultivated in the hotter and damper parts of India.

Part/s Used : Leaves.

Utilization : Betel leaves possess a strong aromatic flavour and used as a masticatory with areca nuts. Also, yield an essential oil.

Active Principles : These are chavicol, chavibetol, allylpyrocatechol, chavibetol acetate, allyl pyrocatechol diacetate (leaves), chavibetol (53.1), chavibetol acetate (15.5), caryophyllene (3.79), allylpyrocatechol diacetate (0.71), camphene (0.48), chavibetol methyl ether (0.48), eugenol (0.32), α-pinene (0.21), β-pinene (0.21), α-limonene (0.14), safrole (0.11), 1,8-cineole (0.04), p-cymene (0.11) and allyl pyrocatechol monoacetate (leaf oil).

References : Ambasta (1986), Bhattacharjee (1998), Chevallier (1996), Kirtikar and Basu (1984), Rastogi and Mehrotra (1993, 1995), Watt (1972).

460. *Piper cubeba* Linn.f.

Syn : Cubeba officinalis Miq.

Family : Piperaceae.

English, Sanskrit and Hindi Names : **Eng. :** Cubebs, Tailed Pepper; **Sans. :** Kakkola, Kankolaka; **Hindi :** Kabab chini.

Distribution : Cultivated to a small extent in India.

Part/s Used : Fruits.

Utilization : Fruits, called cubebs, used as a spice. Oil of cubebs having flavour of camphor and peppermint, used in lozenges and for flavouring bitters, alcoholic and non-alcoholic beverages, baked goods, confectionery, meat and meat products, condiments, cigarettes, relishes, soaps, detergent creams, lotions, perfumes and tobacco.

Active Principles : Oil contains sesquiterpenes cubebin, (2.5%), caryophyllene, cadinene, α- and β-cubebene, copaene, sabinene, α-thujene, β-phellandrene, limonene and acimene.

References : Ambasta (1986), Anonymous (1999), Bhattacharjee (1998), Chevallier (1996), Greenish (1999), Harborne and Baxter (2001), Rastogi and Mehrotra (1990, 1991), Watt (1972).

461. ***Piper longum*** Linn.

Syn : Chavica roxburghii Miq.; *C. sarmentosa* Miq.; *Piper latifolium* Hunter.; *P. sarmentosum* Wall.

Family : Piperaceae.

English, Sanskrit and Hindi Names : **Eng. :** Indian Long Pepper, Piper root; **Sans. :** Chapala, Chanchala, Granthika, Magadhi, Pippali; **Hindi :** Pipal, Pipar, Piplamul, Pipli, Pipulmul.

Distribution : Warmer regions of India. C. Himalaya. Cultivated in Assam, Tamil Nadu and Cherrapunji.

Part/s Used : Fruits.

Utilization : Fruits used as spices and also yield 0.7% essential oil.

Active Principles : Fruits contain piperine (mainly responsible for its sharp taste), piplartine, dihydrostigmasterol, hexadecane.

References : Ambasta (1986), Anonymous (1999), Kirtikar and Basu (1984), Rastogi and Mehrotra (1990,1993), Sinha and Sinha (2001), Smit (2000), Watt (1972).

462. ***Piper mullesua*** Buch.-Ham.

Syn : P. brachystachyam Wall. *ex* Hook.f.

Family : Piperaceae.

English Names : **Eng. :** Pahari Pan, Pahari Pipar.

Distribution : Himalaya: 600-1500m; Khasia Hills: 900-1,500m; Nilgiri Hills at 1500m.

Part/s Used : Leaves.

Utilization : Leaves used as condiments and also yield a volatile oil with an odour akin to lime oil.

Active Principles : Caryophyllene epoxide isolated from essential oil.

References : Ambasta (1986), Mittre (1981), Rastogi and Mehrotra (1991).

463. ***Piper nigrum*** Linn.

Syn : Muldera multinervis Miq.; *M. wightiana* Miq.; *Piper attenuatum* Buch.-Ham. *ex* Miq.; *P. nigrum* var. *trioicum* Cas. DC.; *P. trioicum* Roxb.

Family : Piperaceae.

English, Sanskrit and Hindi Names : **Eng. :** Black Pepper, Common Pepper, Pepper; **Sans.:** Dharmapattana, Hapusha, Katuka, Maricha, Ushana; **Hindi:** Chocamirch, Golmirch, Gulmirch, Habush, Kalimirch, Mirch.

Distribution : Cultivated in the hotter and moist parts of S. India.

Part/s Used : Fruits. Seeds.

Utilization : Fruits used as a condiment and seeds as a spice. Peppers yield an essential oil, called Pepper oil, used in the flavouring of sausages, canned meats, soups, fish, pickles, table sauces, ketchups, baked goods, certain beverages and liquors and also in perfumery.

Active Principles : The acrid taste of herb is due to piperine. Oil contains sabinene, myrcene, limonene, α- and β- pinenes, β-caryophyllene, α-bergamotene, α-humulene, p-cymene, α-selinene, β-bisabolene and camphene. Pipercide isolated from fruits.

References : Ambasta (1986), Chevallier (1996), Greenish (1999), Kirtikar and Basu (1984), Mehtani and Ingle (1999), Rao and Henry (1995), Rastogi and Mehrotra (1990, 1991, 1995), Watt (1972).

464. *Piper peepuloides* Roxb.

Family : Piperaceae.

Hindi Name : **Hindi :** Savali peepul.

Distribution : Tropical Himalaya. Assam and Khasia Hills ascending to 900m.

Utilization : Used as a condiment.

Active Principles : Piptaline, N-isobutyldeca-trans-2-trans-4-diena-mide isolated. Peepuloidin and cinnamolpyrrolidine amide (I) isolated from leaves.

References : Arora (1981), Rastogi and Mehrotra (1990,1993,1995).

465. *Piper retrofractum* Vahl

Syn : Chavica maritima Miq.; *C. officinarum* Miq.; *C. peepuloides* Wight. (not of Roxb.); *Piper arnottianum* Cas.DC.; *P.callosum* Opis.; *P.chaba* Hunter (non-Blume); *P. glabrum* Roxb.; *P.longum* Blume; *P. maritimum* Blume; *P. officinarum* DC.; *P.peepuloides* Wall.

Family : Piperaceae.

English, Sanskrit and Hindi Names : **Eng. :** Cubeb, Java Long Pepper; **Sans. :** Cavika, Chavi, Gandhanakuli, Katuka, Kola; **Hindi :** Chab, Chavi.

Distribution : Cultivated in various parts of India.

Part/s Used : Fruits. Stem.

Utilization : Fruits used as a spice and stem for flavouring curries.

Active Principles : Fruits contain N- isobutyltrideca-13 (3,4-meth-ylene dioxyphenyl) 2,4,12-trienamide (I), filfiline, N – isobuty-leicosa-trans-2, trans-4, cis-8, trienamide (II), piperoctadecalidine, pipereicosalidine- piperine, pipernonaline and guineesine.

References : Ambasta (1986), Anonymous (1999), Kirtikar and Basu (1984), Rastogi and Mehrotra (1991), Tarafder (1983), Watt (1972).

466. *Piper schmidtii* Hook.f.

Family : Piperaceae.

English Name : **Eng.** : Nilgiri Pepper.

Distribution : Assam. W. Ghats. Nilgiris and Palni hills in Tamil Nadu ascending to 1,500m.

Part/s Used : Fruits.

Utilization : Used as a spice and condiment.

Active Principles : Galgravin, friedelin, neolignan-schmiditin, octacosanoic acid, 1-triacontanol, β-sitosterol and its glucoside isolated from fruits.

References : Ambasta (1986), Mittre (1981), Rastogi and Mehrotra (1995).

467. *Piper trichostachyon* DC.

Family : Piperaceae.

Distribution : W. Ghats upto 750m. Mysore. Kerela. Sivagiri hills in Tirunelveli district.

Part/s Used : Spikes. Fruits.

Utilization : Fragrant spikes and fruits constitute the pouched Pepper of Malabar.

References : Ambasta (1986), Rastogi and Mehrotra (1991).

468. *Pistacia khinjuk* Stocks

Syn : *P. chinensis* Bunge ssp. *integerrima* (Steward) Rech.f.; *P. integerrima* Steward *ex* Brandis; *Rhus integerrima* Wall.; *R.kakrasingee* Royle (III).

Family : Anacardiaceae.

Sanskrit and Hindi Names: **Sans.** : Chakra, Ghosha, Kolira, Natangi, Shikhari, Vakra; **Hindi** : Kakra.

Distribution : Hot slopes of the W. Himalaya: 400-2,700m.

Part/s Used : Seed Kernel.

Utilization : Seed kernels used to flavour sweets and cakes. Seed oil aromatic. Galls gave 1.3% essential oil.

Active Principles : Essential oil contains α-pinene (25), camphene (27), dl-limonene (4-5), 1,8-cineole (10), α-terpineol (20), aromadendrene (4-5%) and caprylic acid.

References : Bennett (1987), Chandra (1997), Chopra *et al.* (1956), Kirtikar and Basu (1984), Rastogi and Mehrotra (1995), Watt (1972).

469. *Pistacia lentiscus* Linn.

Family : Anacardiaceae.

English Names: **Eng.** : Mastiche, The Mastic Tree.

Distribution : Native of the Mediterranean region; introduced into India.

Part/s Used : Bark exudate.

Utilization : Used for flavouring alcoholic beverages, cordials and chewing gum.

Active Principles : Resin contains α - and β-masticoresins, a volatile oil (comprising mainly α -pinene), tannins, masticin and mastic acid.

References : Chevallier (1996), Mehtani and Ingle (1999), Rastogi and Mehrotra (1993, 1995), Watt (1972).

470. ***Pistacia vera*** Linn.

Family : Anacardiaceae.

English Name : **Eng.** : Pistachio.

Distribution : Cultivated throughout the Mediterranean area. Introduced from Afghanistan and Iran.

Part/s Used : Kernel.

Utilization : Fruits have a unique flavour, much used as an ingredient of sweetmeats, confectionery and ice-creams.

Active Principles : Nut contains cyanidin-3-O-galactoside, a mixture of 6-alkyl (III, IV) and cis-6-alkenyl (V) salicyclic acids.

References : Ambasta (1986), Harborne and Baxter (2001), Mehtani and Ingle (1999), Rastogi and Mehrotra (1993).

471. ***Pithecellobium angulatum*** Benth.

Family : Mimosaceae.

Distribution : E. Himalaya.

Part/s Used : Flowers.

Utilization : Flowers fragrant.

References : Borthakur (1981), Chopra *et al.* (1980).

472. ***Pittosporum undulatum*** Vent.

Family : Pittosporaceae.

Part/s Used : Flowers.

Utilization : Flowers yield an essential oil having a jasmine-like odour.

Reference : Ambasta (1986).

473. ***Pluchea indica*** Less.

Syn : *Baccharis indica* Linn.; *Conyza corymbosa* Roxb.; *C. indica* Miq.; *Pluchea foliolosa* DC.

Family : Asteraceae.

Distribution : Salt marshes of the Sunderbans.

Part / s Used : Leaves.

Utilization : Aromatic leaves used as a flavouring ingredient.

Active Principles : Plucheosides A and B, linalyl glucoside, linalyl apiosyl glucoside and 9-hydroxylinalyl glucoside isolated from aerial parts.

References : Ambasta (1986), Kirtikar and Basu (1984), Rastogi and Mehrotra (1995), Watt (1972).

474. ***Pogostemon cablin*** (Blanco) Benth.

Syn : *Origanum indicum* Roth.; *P. heyneanus* Benth.; *P. patchouli* Hook.f. (non Pellet.) var. *suavis* Hook.f.

Family : Lamiaceae.

English and Hindi Names: **Eng.** : Patchouli; **Hindi** : Pacholi, Peholi.

Distribution : Kanara. W. Ghats from Nilgiris southwards, wild and cultivated.

Part / s Used : Leaves.

Utilization : Fresh leaves used for flavouring many food products; dried ones for scenting wardrobes and woolens. Also an essential oil, Patchouli oil extracted from leaves and used for scenting soaps, tobacco and incenses and as an excellent masking agent in depilatory creams. It yields an excellent 'attar' on blending with sandalwood oil.

Active Principles : These are seychellene, epiguaipyridine, patchouli pyridine, dhelwangin, δ-patchoulene, patchouli alcohol (32.77), α-bulnesene (14.85) and α-patchoulene (12.74%) (essential oil), 30-40% (-)-patchoulol and nor-patchoulenol (leaves).

References : Ambasta (1986), Bennett (1987), Chevallier (1996), Chopra *et al.* (1956), Harborne and Baxter (2001), Kumar and Singh (2000), Rastogi and Mehrotra (1990, 1991, 1993, 1995), Watt (1972).

475. ***Polianthes tuberosa*** Linn.

Family : Liliaceae.

English, Sanskrit and Hindi Names : **Eng.** : Tuberose; **Sans.** : Rajnigandha; **Hindi** : Gulcheri, Gulshabbo.

Distribution : Widely cultivated in India.

Part / s Used : Flowers.

Utilization : Fragrant flowers yield an essential oil, Tuberose oil, much valued in high grade perfumery.

Active Principles : Tuberose oil yields benzyl alcohol, butyric acid, eugenol, farnesol, geraniol, methyl anthrinilate, methyl benzoate, methyl salicylate, nerol and phenyl acetic acid.

References : Ambasta (1986), Bhattacharjee (1998), Kirtikar and Basu (1984), Rastogi and Mehrotra (1991), Watt (1972).

476. ***Polygonum hydropiper*** Linn.

Syn : P. flaccidum Meissn.; *P. mite* Wall.

Family : Polygonaceae.

English Names: **Eng.** : Arsesmart, Biting Pepper, Pepper-wort, Smartweed, Water Pepper.

Distribution : Throughout India.

Part/s Used : Young shoots. Leaves.

Utilization : Used for flavouring raw fish, and also as spice.

Active Principles : Leaves contain polygonal, hemiacetal-isodrime-ninol, polygodial, isopolygodial, confertifolin, warburganol, drimenol and polygonic acid.

References : Ambasta (1986), Bennett (1987), Bhattacharjee (1998), Chandra (1997), Kirtikar and Basu (1984), Mehtani and Ingle (1999), Rastogi and Mehrotra (1990, 1991, 1993, 1995), Watt (1972).

477. ***Polygonum microcephalum*** D.Don.

Syn : P. ciliatum Ham.; *P. staticiflorum* Wall.; *P. strigosum* Herb. Ham.

Family : Polygonaceae.

Distribution : Sikkim. Khasia mountains. Hills of Assam and Nagaland.

Part/s Used : Young Tops.

Utilization : Young tops used in Assam for flavouring vegetables.

References : Ambasta (1986), Mehtani and Ingle (1999), Watt (1972).

478. ***Polyscias fruticosa*** (Linn.) Harms.

Syn : Nothopanax fruticosum Miq.; *Panax fruticosum* Linn.

Family : Araliaceae.

Distribution : Cultivated in gardens all over India.

Part/s Used : Leaves.

Utilization : Leaves used in the flavouring.

References : Ambasta (1986), Bennett (1987), Mehtani and Ingle (1999).

479. ***Poncirus trifoliata*** (L.) Rafin.

Syn : Citrus trifoliata Linn.

Family : Rutaceae.

English Name: **Eng.** : Trifoliate Orange.

Distribution : Introduced into India mainly Assam.

Part/s Used : Fruits (juice, peel, mesocarp).

Utilization : Peel makes excellent spicy flavouring for cakes. Syrup made from the fruit juice also used as a flavouring agent.

Active Principles: 7-geranyloxycoumarin, 6-methoxy-7-geranylo-xycoumarin, bergapten and imperatorin isolated from fruits.

References : Ambasta (1986), Mehtani and Ingle (1999), Rastogi and Mehrotra (1990, 1991).

480.*Porana paniculata* Roxb.

Family : Convolvulaceae.

English Name : **Eng. :** Bridal Creeper.

Part/s Used : Flowers.

Utilization : Flowers have faint lavender like fragrance and used in bouquets.

Reference : Ambasta (1986).

481.*Pothomorphe subpeltata* (Willd.) Miq.

Syn : Piper subpeltatum Willd.

Family : Piperaceae.

Distribution : W. Ghats.

Part/s Used : Flower. Young leaves.

Utilization : Both flowers and young leaves used for flavouring fish curry.

Reference : Mehtani and Ingle (1999).

482.*Prangos pabularia* Lindl.

Family : Apiaceae.

Hindi Name : **Hindi :** Komal.

Distribution : Common in Dras in Ladakh and Zanskar range (W. Himalaya).

Part/s Used : Roots.

Utilization : Roots burnt as incense, also yield an essential oil

Active Principles : These are osthol, merancin, psoralen, xanthotoxol, xanthotoxin, imperatorin, prangenin, pabulari-none, impera-torin, isooxypeucedanin, prangenin [(S) heraclenin] (roots), camphene, limonene, car-3-ene, phellandrene, terpinolene, camphor, carvone, β- and γ- elemenes, β-caryophyllene, β-selinene, citronellyl acetate, geranyl acetate, eudesmol, elemol, myrcene, α-pinene, osthol and imperatorin (essential oil).

References : Ambasta (1986), Kaul (1997), Rastogi and Mehrotra (1990, 1991, 1993), Watt (1972).

483. ***Premna latifolia*** Roxb.

Syn : P. mucronata Roxb.

Family : Verbenaceae.

Hindi Names : **Hindi :** Agnium, Bakar, Bakarcha, Basot, Basota, Jhatela.

Distribution : Sub-Himalayan Tract. Khasia mountains, ascending to 1,700m.

Part/s Used : Leaves.

Utilization : Leaves used for flavouring curries.

Active Principles : Chief ingredients extracted from leaves are furanoid – 2,4-di (3',4'-methylenedioxyphenyloxy) furan (prem-nalatin), β-sitosterol, 5-hydroxy-4'-methoxyflavone-7-O-arabino-pyranosyl rhamnoside and 5-hydroxy-4'-methoxyflavone-7-O-trioside.

References : Kirtikar and Basu (1984), Mittre (1981), Rastogi and Mehrotra (1993), Watt (1972).

484. ***Prunus amygdalus*** Batsch (Pl. 10B)

Syn : Amygdalus communis Linn.; *P. communis* Fritsch; *P. dulcis* (Mill.) D.A. Webb.

Family : Rosaceae.

English, Sanskrit and Hindi Names : **Eng. :** Almond Tree; **Sans. :** Badama, Netropamaphala, Suphala, Vatada; **Hindi :** Badam.

Distribution : Cultivated in Kashmir and the Punjab.

Part/s Used : Kernels.

Utilization : Almond oil extracted from sweet as well as bitter almonds used in baking, confectionery and cosmetics. Bitter almonds also used for flavouring.

Active Principles : These are prunasin, daucosterin and sitosterol (bitter seeds), daucosterin and sitosterol (sweet almonds), benzaldehyde and hydrocyanic (prussic acid) acid (essential oil of bitter almonds), olein and linolein (almond oil). Bitter almonds differ from sweet in containing a colourless crystalline glucoside, amygdalin.

References : Ambasta (1986), Greenish (1999), Harborne and Baxter (2001), Kirtikar and Basu (1984), Mehtani and Ingle (1999), Rastogi and Mehrotra (1991,1995), Watt (1972).

485. ***Prunus armeniaca*** Linn.

Family : Rosaceae.

English and Hindi Names : **Eng. :** Common Apricot, Mishmus, Moon of the Faithful; **Hindi :** Chilu, Chuari, Jaldaru, Khubani, Kushmiaru, Zardalu.

Distribution : Commonly cultivated in the N.W.Himalaya and Punjab plains.

Part/s Used : Kernels.

Utilization : Fatty oil extracted from kernels used in cosmetic industry. Essential oil possesses a pleasant flavour of bitter almonds.

Active Principles : Seeds contain upto 8% amygdalin, 24-dehydrocholesterol, cholesterol, estrone, α-estradiol (estradiol-β-17-ol), myristic, palmitic, stearic, oleic and linoleic acids and β-sitosterol.

References : Ambasta (1986), Chevallier (1996), Kirtikar and Basu (1984), Rastogi and Mehrotra (1991), Watt (1972).

486.*Prunus cerasus* Linn.

Syn : *Cerasus caproniana* DC.; *C. vulgaris* Miller.

Family : Rosaceae.

English and Hindi Names : **Eng.** : Cherry Tree, Dwarf Cherry, Red Cherry, Sour Cherry, Wilol Cherry; **Hindi** : Alubalu.

Distribution : Cultivated in Himalaya and Punjab.

Part/s Used : Kernels.

Utilization : Fatty oil extracted from kernels used in cosmetic industries and as a salad or culinary oil. Kernels also used for flavouring liquors.

Active Principles : Fruits yield quercetin-3-O-rutinosyl-7,3'-O-biglucoside, kaempferol-3-O-rutinosyl-4'-di-O-glucoside and two isomers of quercetin-3-O-rutinosyl-4'-di-O-glucoside.

References : Ambasta (1986), Rastogi and Mehrotra (1993), Watt (1972).

487.*Prunus domestica* Linn.

Syn : *P. aloocha* Royle; *P. bokhariensis* Royle; *P. communis* Huds.; *P. insititia* Linn. (the Bullace).

Family : Rosaceae.

English and Hindi Names : **Eng.** : Alucha, Common Plum Tree; **Hindi** : Alubukhara, Alucha.

Distribution : Common throughout the N.W. Himalaya.

Part/s Used : Fruits. Kernels.

Utilization : Dried plums (Prunes) are often added to cathartic decoctions for improvement of flavour. Kernels yield plum oil with applications similar to that of almond oil.

Active Principles : Fruits contain galacturonic acid, arabinose, galactose, xylose and rhamnose.

References : Ambasta (1986), Greenish (1999), Kirtikar and Basu (1984), Rastogi and Mehrotra (1995), Watt (1972).

488. *Prunus laurocerasus* Linn.

Syn : Laurocerasus officinalis M.Roem.

Family : Rosaceae.

English Name : **Eng. :** Cherry Laurel.

Part/s Used : Leaves.

Utilization : Macerated leaves in water (Cherry Laurel Water) used as a substitute for flavour of bitter almonds. Essential oil, known as Oil of Cherry Laurel, obtained by redistilling the distillation water.

Active Principles : Chery Laurel leaves contain mainly Prulaurasin.

References : Ambasta (1986), Greenish (1999).

489. *Prunus mahaleb* Linn.

Family : Rosaceae.

English and Sanskrit Names : **Eng. :** Mahaleb Cherry; **Sans. :** Priyunger.

Part/s Used : Fruits. Leaves.

Utilization : Both fruits and leaves used for flavouring liquors and in perfumery. Also, leaves used for flavouring sauces.

Active Principles : Kaempferol (0.74%) from leaves.

References : Ambasta (1986), Kirtikar and Basu (1984), Rastogi and Mehrotra (1990), Watt (1972).

490. *Prunus persica* (L.) Batsch

Syn : Amygdalus collinus Wall.; *A. persica* Linn.; *P. saligna* Royle; *P. vulgaris* Miller.

Family : Rosaceae.

English and Hindi Names : **Eng. :** Nectarine, Peach Tree; **Hindi :** Aru, Shaftalu.

Distribution : Commonly cultivated in the Himalaya. Plains of N.India. Manipur.

Part/s Used : Fruits. Kernels.

Utilization : Peaches are used in beverages. Kernels yield a fatty oil, called Peach Kernel oil with applications similar to bitter almond oil.

Active Principles : Flavour constituents of Peaches are due to undecalactone. 1-naphthylacetic acid and 2-(1-naphthyl-acetamide, oleic acid (1.5) and linoleic acid (0.5%) (fruits). Essential oil yields γ-hexalactone, γ-octalactone, δ-octalactone, γ-nonalactone, γ-decalactone, δ-decalactone, γ-dodecalactone and δ-dodecalactone.

References : Ambasta (1986), Harborne and Baxter (2001), Kirtikar and Basu (1984), Rastogi and Mehrotra (1991), Watt (1972).

491. ***Psidium guajava*** Linn.

Family : Myrtaceae.

English, Sanskrit and Hindi Names : **Eng.** : Common Guava Tree; **Sans.** : Dridhabija, Mansala, Perala, Vastula; **Hindi** : Amrud, Amrut, Safed Safari.

Distribution : Cultivated and naturalized throughout India.

Part/s Used : Fruits. Leaves.

Utilization : Leaves contain an essential oil, used as a flavouring agent. The pulp pleasantly fragrant.

Active Principles : Essential oil isolated from leaves contain myrcene (0.99), dl-limonene (8.01), caryophyllene (7.7), caryophyllene oxide (37.5), eugenol (13.0) and 1,8-cineole (2.7%). Fruits yield hexyl acetate, ethyl decanoate, β-caryophyllene, α-humulene, α-selinene, myrcene, benzaldehyde, α –copaene and oleanoilc acid.

References : Ambasta (1986), Kirtikar and Basu (1984), Mehtani and Ingle (1999), Raju (2000), Rastogi and Mehrotra (1990, 1991, 1995), Watt (1972).

492. ***Pteris ensiformis*** Burm.f.

Family : Pteridaceae.

Part/s Used : Fronds.

Utilization : Young fronds eaten as a flavouring.

References : Ambasta (1986), Singh and Viswanathan (1996).

493. ***Pterocarpus marsupium*** Roxb.

Syn : *P. bilobus* Roxb.

Family : Papilionaceae.

English, Sanskrit and Hindi Names : **Eng.** : Indian Kino Tree, Malabar Kino Tree; **Sans.** : Asana, Bijavriksha, Mahakutaj, Nilaka, Pitashala, Sauri; **Hindi** : Banda, Bija, Bijasal, Hiradokhi, Piasal, Pitshola.

Distribution : C. and W. Peninsular India. Mixed deciduous forests.

Part/s Used : Flowers.

Utilization : Flowers sweetly fragrant.

Active Principles : These are 4,6,4'-trihydroxyaurone-6-O-rhamnopyranoside (I) and 4,6,4'-trihydroxy-7-methylaurone-4-O-rhamnopyrnoside (II).

References : Anonymous (1999), Bhattacharjee (1998), Chevallier (1996), Jain (1968), Kirtikar and Basu (1984), Rastogi and Mehrotra (1995), Sahni (1998), Watt (1972).

494. *Punica granatum* Linn. (Pl. 10C)

Family : Punicaceae.

English, Sanskrit and Hindi Names : **Eng.** : Pomegranate Tree; **Sans.** : Bijapura, Dadima, Karaka, Milapatra, Suphala; **Hindi** : Anar, Dhalim, Dharimb.

Distribution : N.W. India, but cultivated in the warmer parts of the Temperate regions.

Part/s Used : Dried Seeds.

Utilization : Dried seeds, 'Anardana', used as spices and condiments.

Active Principles : The seeds contain steroidal estrogens.

References : Ambasta (1986), Anonymous (1999), Bhattacharjee (1998), Chevallier (1996), Greenish (1999), Kirtikar & Basu (1984), Negi & Gaur (1994), Raju (2000), Rastogi & Mehrotra (1991), Singh (1999).

495. *Pycnanthemum virginianum* Durand and Jackson

Syn : *P. lanceolatum* Pursh

Family : Lamiaceae.

English Name : **Eng.** : Virginia Mountain Mint.

Distribution : Grown in gardens.

Part/s Used : Flowers.

Utilization : Flowers used to flavour meat and fish.

Active Principles : Iso-menthone, limonene and pulegone isolated from flowers.

Reference : Mehtani and Ingle (1999).

496. *Pyrus pashia* Buch.-Ham. *ex* D.Don

Syn : *P. kumaoni* Decne. *ex* Hook.f.; *P. nepalensis* Ham.; *P. variolosa* Wall.; *P. verruculosa* Bertol.

Family : Rosaceae.

Hindi Name : **Hindi** : Mehal.

Distribution : Temperate Himalaya. Khasia mountains.

Part/s Used : Fruits.

Utilization : Flavour and taste resemble that of pear.

Active Principles : β-sitosterol and its glucoside isolated.

References : Arora (1981), Bennett (1987), Rastogi and Mehrotra (1990, 1991), Sahni (1998), Watt (1972).

497. *Ramalina calicaris* (Linn.) Rohl.

Family : Usneaceae.

Part/s Used : Lichen Thallus.

Utilization : Used as a spice and in perfumery.

References : Ambasta (1986), Upreti *et al.* (2002).

498. *Ramalina farinacea* (Linn.) Ach.

Family: Usneaceae.

Distribution : Manipur. Tamilnadu. Uttar Pradesh. W. Bengal. Nilgiris.

Part/s Used : Lichen Thallus.

Utilization : Used as spices and also in perfumery and cosmetics.

Active Principles : Evernic acid, norstictic acid, obtusatic acid, sekikaic acid, (+) usnic acid and mannitol isolated from thallus.

References : Ambasta (1986), Mudgal and Hajra (1997), Rastogi and Mehrotra, Upreti *et al.* (2002).

499. *Ramalina fraxinea* (Linn.) Ach.

Family : Usneaceae.

Distribution : Coromandal. H. P. Manipur. W. Bengal.

Part/s Used : Lichen Thallus.

Utilization : Used as spices and also in perfumery and cosmetics.

Active Principles : Usnic acid and arabitol.

References : Ambasta (1986), Mudgal and Hajra (1997), Upreti *et al.* (2002).

500. *Ramalina subcomplanata* (Nyl.) Kashiwadani

Family : Usneaceae.

Distribution : Assam. H.P. Nagaland. Uttar Pradesh.

Part/s Used : Lichen.

Utilization : Used as spices to flavour non-vegetarian preparations, pulses and other vegetables.

Active Principles : Salicinic acid and Usnic acid.

References : Mudgal and Hajra (1997), Upreti *et al.* (2002).

501. *Raphia vinifera* Beauv.

Family: Arecaceae.

Part/s Used : Fruit Pulp.

Utilization : Yellow oily pulp used as a bitter flavour.

Reference : Ambasta (1986).

502. *Remirea maritima* Aubl.

Syn : Cyperus pedunculatus (R.Br.) Kern

Family : Cyperaceae.

Part/s Used : Roots.

Utilization : Roots aromatic.

Active Principles : Flavonone quinone – remerin isolated from roots.

References : Ambasta (1986), Chopra *et al.* (1956), Rastogi and Mehrotra (1991).

503. *Remusatia vivipara* Schott

Syn : *Arum viviparum* Roxb.

Family : Araceae.

Distribution : Sub-tropical Himalaya. Khasia Hills. Chota Nagpur. Mysore.

Part/s Used : Roots.

Utilization : Roots aromatic.

References : Ambasta (1986), Kirtikar and Basu (1984).

504. *Reseda odorata* Linn.

Family: Resedaceae.

English Name: **Eng.** : Mignonette.

Distribution : Widely cultivated in Indian gardens.

Part/s Used : Flowers.

Utilization : Mignonette oil extracted from flowers finds application in high grade perfumery.

Active Principles : 2-hydroxybenzylamine, 2-(α-L-rhamnopyr-anosy-loxy) and a new aminoacid –2(S), 4(R)4-(β-D-galactopyra-nosy-loxy)- 4-isobutyl glutamic acid (I) – isolated from flowers.

Reference : Ambasta (1986), Kirtikar and Basu (1984),Rastogi and Mehrotra (1991), Watt (1972).

505. *Rhamnus purshiana* DC.

Family : Rhamnaceae.

Part/s Used : Bark.

Utilization : Debittered extract of bark employed for flavouring baked food, ice-cream and soft drinks.

Active Principles : Bark contains anthracene compounds, anthraquinones, emodin, oxanthrone, aloe-emodin, chrysophenol and dianthrones.

Reference : Bhattacharjee (1998).

506. *Rheum rhaponticum* Linn.

Family : Polygonaceae.

English Names: **Eng.** : Rhapontic Rhubarb, Rhei rhapontici radix, Pie-plant, Garden rhubarb.

Distribution : Cultivated in Nilgiris, Khasia Hills and hills of W. Bengal.

Part/s Used : Petioles.

Utilization : Pleasantly acidic petioles used in sauces and pies for flavour.

Active Principles : These are citric, malic and oxalic acids and 3, 3', 4', 5-tetrahydroxystilbene-3-glucoside.

References : Ambasta (1986), Mehtani and Ingle (1999).

507.*Rhododendron anthopogon* D.Don

Syn : *Azalea lapponica*; *R. aromaticum* Wall.; *R. haemonium* Balf.f. & Cooper; *Osmothamnus fragrans* DC.; *O. pallidus* DC.

Family : Ericaceae.

Distribution : Sub-Alpine area of H.P: 2,800-3,800m. Alpine Himalaya: 3,700-5,400m.

Part/s Used : Leaves.

Utilization : Leaves strongly scented and used as an incense.

Active Principles : Quercetin-3-O-α-L-rhamnopyranoside, kaempferol, its methyl ether and 3-O-glucoside isolated from leaves.

References : Ambasta (1986), Bennett (1987), Bhattacharjee (1998), Dam and Hajra (1981), Kirtikar and Basu (1984), Rastogi and Mehrotra (1995), Rawat and Pangtey (1987), Saklani and Jain (1994), Sood *et al.* (2001), Watt (1972).

508.*Rhododendron lepidotum* Wall. *ex* G.Don

Syn : *R. cremnastes* Balf.f. and Farrer; *R. elaeagnoides*; *R. obovatum* Hook.f. *R. salignum* Hook.f.

Family : Ericaceae.

Distribution : Temperate and Alpine Himalaya: 2,700-5,000m.

Part/s Used : Leaves.

Utilization : Volatile oil extracted from leaves used in low grade perfumes and incenses.

References : Ambasta (1986), Bennett (1987), Bhattacharjee (1998), Kirtikar and Basu (1984), Watt (1972).

509.*Rhododendron setosum* D.Don

Syn : *R. anthopogon* Wall.

Family : Ericaceae.

Distribution : Sikkim: 4,350-5,350m

Part/s Used : Leaves.

Utilization : Oil from leaves finds application in perfumery and incenses.

References : Ambasta (1986), Kirtikar and Basu (1984), Watt (1972).

510. *Rhodomyrtus parviflora* Alston

Syn : R. tomentosa Wight in part, non Hasck.; *Myrtus canescens* Lour; *M. tomentosa* Ait.

Family : Myrtaceae.

English Names: **Eng. :** Downy Rosemyrtle, Hill-Gooseberry.

Distribution : Higher mountains of S. India.

Part/s Used : Fruits.

Utilization : Fruit pulp aromatic and similar to apple jelly.

References : Ambasta (1986), Mehtani and Ingle (1999), Watt (1972).

511. *Rhus semialata* Murray

Syn : R.amela Don; *R.buckiamela* Roxb.; *R.chinensis* Mill.; *R.javanica* Linn.; *R. javanica* var. *roxburghii* (DC.) Rehder and Wilson.

Family : Anacardiaceae.

Distribution : Outer Himalayan ranges:1,000-2,400m.

Part/s Used : Flowers. Fruits.

Utilization : Flowers fragrant.

Active Principles : Triterpene-semialatic acid isolated.

References : Bennett (1987), Kirtikar and Basu (1984), Negi and Gaur (1994), Rastogi and Mehrotra (1990), Watt (1972).

512. *Rhynchostylis retusa* Bl.

Family : Orchidaceae.

English Name : **Eng. :** Cats Tail Orchid.

Distribution : Chhota Nagpur. W. Bengal. Orissa. W. Ghats Malabar Coast. H.P.

Part/s Used : Flowers.

Utilization : Flowers pleasantly odorous.

Reference : Bhattacharjee (1998).

513. *Ribes nigrum* Linn

Family : Grossulariaceae.

English Names: **Eng. :** The Black Currant, Quinsyberry, Squinancy.

Distribution : Temperate W. Himalaya.

Part/s Used : Buds. Fruits.

Utilization : Essential oil from buds used for flavouring ice-cream, yoghurt and alcoholic beverages. Fruits aromatic and used in sauces and pies, juices, syrups, wines, jams and jellies.

Active Principles : Bud essential oil contains γ-elemene, γ-muurolene, ledene, bicyclogermacrene, germacrene B, trans-and cis-limonene, 1,2-epoxide, linalyl and neryl acetates, spathulenol and isospathulenol.

References : Ambasta (1986), Chevallier (1996), Kirtikar and Basu (1984), Mehtani and Ingle (1999), Rastogi and Mehrotra (1991, 1995), Watt (1972).

514.*Rimelia reticulata* (Taylor) Hale and Fletcher

Family : Parmeliaceae.

Part/s Used : Lichen Thallus.

Utilization : Used as spices.

Reference : Upreti *et al.* (2002).

515.*Robinia pseudoacacia* Linn.

Family : Papilionaceae.

English Names: **Eng. :** Black locust, False-acacia, Robinia, The North American Locust Acacia.

Distribution : Grown as ornamental tree in Indian gardens.

Part/s Used : Flowers.

Utilization : Essential oil from flowers used as a flavouring and also finds application in the preparation of aromatic waters and 'sherbets'.

Active Principles : These are phenylethyl alcohol (12.07), linalool (3.11), geraniol (2.14), nerolidol (0.89), nonalactone (0.86), teresantalol (0.21), ethyl hexanoate (0.20), benzyl alcohol (0.17) and ethyl octanoate (0.13%) (flower essential oil).

References : Ambasta (1986), Mehtani and Ingle (1999), Rastogi and Mehrotra (1990, 1991), Watt (1972).

516.*Roccella belangeriana*

Family : Roccellaceae.

Part/s Used : Lichen Thallus.

Utilization : Used as spices and for flavouring non-vegetarian preparations, pulses and other vegetables.

Reference : Mudgal and Hajra (1997).

517.*Roccella montagnii* Bel.

Family : Roccellaceae.

Distribution : Karnataka. Coastal regions of Orissa, Andhra Pradesh. Pondicherry.

Part/s Used : Lichen Thallus.

Utilization : Used as spices for flavouring non-vegetarian preparations, pulses and other vegetables.

Active Principles : β-carotene, γ-carotene and ergasterol isolated.

References : Chopra *et al.* (1980), Mudgal and Hajra (1997), Rastogi and Mehrotra (1990).

518. ***Rosa bourboniana*** Desportes

Syn : R. chinensis Jacq.; *R. fragrans* Redoute; *R. indica* Linn.; *R. sinica* Linn.

Family : Rosaceae.

English and Hindi Names: **Eng.** : Bourbon Rosa; **Hindi** : Baramasi, Cheeniagulab, Desi gulab.

Distribution : Cultivated in Indian gardens.

Part/s Used : Petals.

Utilization : The flower petals employed in the formulation of rose water and `gulkand'. Essential oil also isolated from the flowers.

Active Principles : These are gallic acid (flowers), linalool acetate (14.98), limonene (12.07), methoxymethyl benzene (9.88), citronellol (4.82), β-caryophyllene (4.55), hexanol acetate (3.98), linalool (3.18) and hexanol (3.17%) (flower essential oil).

References : Ambasta (1986), Bhattacharjee (1998), Chopra *et al.* (1980), Kirtikar and Basu (1984), Rastogi and Mehrotra (1991, 1995), Watt (1972).

519. ***Rosa brunonii*** Lindl.

Syn : R. glandulifera Herb. Roxb.; *R. lindleyi* Herb. Wall.; *R. moschata* Hook.f., non Mill. *Nec Herrm*; *R.m.* var. *nepalensis* Lindl.; *R. pubescens* Roxb.; *R. recurva* Roxb.

Family : Rosaceae.

English, Sanskrit and Hindi Names: **Eng.** : Himalayan Musk Rose, The Noisette rose; **Sans.** : Sewati; **Hindi** : Kuji, Kunja, Karer, Kwiala.

Distribution : Himalaya from Kashmir to Bhutan, Assam. Also, in Garhwal Himalaya upto 2,000m.

Part/s Used : Flowers.

Utilization : Petals used for the preparation of rose water. An otto (Attar) much valued in perfumery is also extracted from the flowers.

References : Ambasta (1986), Bennett (1987), Bhattacharjee (1998), Raju (2000), Watt (1972).

520. ***Rosa centifolia*** Linn.

Family : Rosaceae.

English, Sanskrit and Hindi Names: **Eng.** : Cabbage Rose, Hundred Leaved Rose, Pale Rose, Provence Rose; **Sans.** : Atikesara, Bhadrataruni, Devataruni, Kharva; **Hindi** : Gulab, Gulabzurdi.

Distribution : Cultivated in India.

Part/s Used : Flowers.

Utilization : Blossoms used for scenting tea and as the source of rose water and otto (`attar') of rose. Floral wax, obtained as a by product of rose perfume, finds application for manufacturing soap. Essential oil also obtained.

Active Principles : Geraniol, nerol, rodinol, phenylethanol and linalool isolated from essential oil.

References : Ambasta (1986), Bhattacharjee (1998), Kirtikar and Basu (1984), Rastogi and Mehrotra (1993), Watt (1972).

521. ***Rosa damascena*** Mill.

Family : Rosaceae.

English, Sanskrit and Hindi Names : **Eng.** : Bussora Rose, Damask Rose, Persian Rose, Summer Damask Rose; **Sans.** : Atimanjula, Lakshapushpa, Mahakumari, Shatapatri; **Hindi** : Bussorah, Fasli Gulab, Gulab, Sudbarg.

Distribution : Widely grown in gardens in India.

Part/s Used : Flowers.

Utilization : Flowers are the source of rose water and otto (`Attar') of rose; and considered the most important and costliest among perfumery. Also, yields comparatively large quantity of volatile oil, Rose oil of Bulgaria, used to flavour snuff, chewing tobacco, fruit flavours, soft drinks, alcoholic liquors and confectionery.

Active Principles : Aromatic constituents of rose oil (Otto of rose) are rhodinol and geranyl acetate. Cyanidin-3,5-diglucoside isolated from petals.

References : Ambasta (1986), Bhakuni (2000), Kapur *et al.* (1996), Kirtikar and Basu (1984), Kumar and Singh (2000), Mehtani and Ingle (1999), Rastogi and Mehrotra (1990, 1991, 1993, 1995), Watt (1972).

522. ***Rosa macrophylla*** Lindl.

Syn. : *Gulellmi waldemarii* Klotssch; *R. hoffmeisteri* Klotssch; *R. hookeriana* Bertol.; *R. macrophylla* var. *hookeriana* Hook.f.

Family : Rosaceae.

Hindi Name : **Hindi** : Ban-gulab.

Distribution : Temperate Himalaya: 2,650-3,500m.

Part/s Used : Flowers.

Utilization : Sweetly scented flowers find their principle use in perfumery and for making rose-water.

References : Ambasta (1986), Bennett (1987), Raju (2000), Watt (1972).

523. *Rosa rubra* Blackw. (Pl. 10D)

Syn : R. gallica Linn.

Family : Rosaceae.

English Names: **Eng. :** French rose, Red officinal rose, Red rose, Rose.

Distribution : Grown in India.

Part / s Used : Flowers.

Utilization : Petals used for making rose water. Essential oil from the flowers used in confectionery.

Active Principles : Chief ingredients are β-Phenethyl-β-D-glucopyranoside, linalool, citronellol, nerol, geraniol, gallicosides A,B and C, citronellol, 2-phenylethanol (flowers) and digeranyl ether (oil).

References : Ambasta (1986), Chevallier (1996), Greenish (1999), Kirtikar and Basu (1984), Rastogi and Mehrotra (1991, 1993), Watt (1972).

524. *Rosa sericea* Lindl.

Syn : R. tetrapetala Royle; *R. wallichii* Trattin.

Family : Rosaceae.

Distribution : Temperate Himalaya: 3,000-4,650m.

Part / s Used : Flowers. Leaves.

Utilization : Both flowers and leaves are fragrant.

References : Ambasta (1986), Bhattacharjee (1998), Negi and Gaur (1994), Watt (1972).

525. *Rosa webbiana* Wallich *ex* Royle

Syn : R. pimpinellifolia Hook.f. & T.

Family : Rosaceae.

English Name : **Eng. :** Wild Rose.

Distribution : Dry inner Himalaya – Kashmir: 1,000-4,500m.

Part / s Used : Flowers.

Utilization : Flowers used for the preparation of rose water.

References : Singh *et al.* (1996), Watt (1972).

526. *Rosmarinus officinalis* Linn.

Family : Lamiaceae.

English and Hindi Names : **Eng. :** Rosemary; **Hindi :** Rusmari.

Distribution : Grown in the Temperate Himalaya and Nilgiris.

Part / s Used : Leaves. Twigs. Flowering Tops.

Utilization : The leaves employed as a condiment and the fresh tender tops used for garnishing and flavouring soft drinks, pickles, soups, etc. Leaves, twigs and flowering tops yield an essential oil, called

Rosemary oil, used in cheap perfumery, cosmetics, room sprays, inhalants, flavouring meats, sausages, soups, table sauces, alcoholic and non-alcoholic beverages, frozen dairy desserts, candy and baked goods, gelatins and puddings.

Active Principles : These are α-and β- pinenes, camphene, sabinene, limonene, 1,8-cineole, α-cymene, p-cymene, linalool, camphor, bornyl acetate, borneol, caryophyllene, cadinene (essential oil; (0.34%), verbenone (12.3), camphor (11.3), bornyl acetate (7.6), limonene (7.1%), camphor (14.9), α-pinene (9.3), 1,8-cineole (9.0%) of leaves (essential oil (0.40%), and rosamaricine (0.33%) (leaves).

References : Ambasta (1986), Bhattacharjee (1998), Chevallier (1996), Chopra *et al.* (1960), Kumar and Singh (2000), Mehtani and Ingle (1999), Rastogi and Mehrotra (1990, 1991, 1993), Sharma and Sood (1997), Watt (1972).

527.*Rubus ellipticus* Sm.

Syn : R. affinis Madden; *R.denudata* Wall.; *R. flavus* Ham.; *R. gowryphul* Roxb.; *R. hirtus* Roxb.; *R. rotundifolius* Wall; *R. sessiflorus* Miq.; *R. wallichianus* W. & Arn.

Family : Rosaceae.

Hindi Names: **Hindi :** Hinsalu, Anchhu.

Distribution : Temperate and Sub-Tropical Himalaya. Khasia Hills. W. Peninsula.

Part/s Used : Fruits.

Utilization : Ripe fruits have flavour and taste like raspberry.

Active Principles : Main constituents are 3β-hydroxy-urs-12,18-dien-28-oic acid-3-O-β-D-glucopyranosyl (1 to 4)- α-L-arabinopy-ranoside, acuminatic acid, tormentic acid and β-sitosterol.

References : Ambasta (1986), Negi and Gaur (1994), Raju (2000), Rastogi and Mehrotra (1995), Watt (1972).

528.*Rubus nepalensis* (Hook.f.) Kuntze

Syn : R. nutaniflorus Hara; *R. nutaniflorus* var. *nepalensis* (Hook.f.) Balak.; *R. nutans* var. *nepalensis* Hook.f.; *R. nutans* Wall. *ex* G.Don.

Family : Rosaceae.

Part/s Used : Fruits.

Utilization : Ripe fruits have an agreeable flavour and taste-like raspberry.

References : Bennett (1987), Negi and Pant (1990).

529. *Rubus niveus* Thunb.

Syn : *R. albescens* Roxb.; *R. gracilis* Roxb.; *R. lasiocarpus* Hook.f. in part.; *R. lasiocarpus* Smith; R.l. var. *membranaceous* Hook.f.; R.l. var. *pauciflorus* (Wall.) Hook.f.; *R. mysorensis* Heyne; *R. niveus* var. *pauciflorus* (Wall.)Deb.

Family : Rosaceae.

English and Hindi Names: **Eng.** : Mahabaleshwar Raspberry, Mysore Raspberry; **Hindi** : Kala hinsalu, Kalianchhi.

Distribution : Temperate Himalaya-Kashmir to Bhutan: 2,000-3,350m. Deccan Peninsula.

Part/s Used : Fruits.

Utilization : Fruits employed for flavouring ice-cream, 'sherbet' and fruit juices.

Active Principles : Raspberry ketone, ionone and geraniol isolated from fruits.

References : Ambasta (1986), Bennett (1987), Mehtani and Ingle (1999), Raju (2000), Watt (1972).

530. *Rumex hastatus* D.Don

Family : Polygonaceae.

Distribution : N.W. Himalaya: 350-2,700m.

Part/s Used : Leaves.

Utilization : Leaves used as condiments.

Active Principles : Anthraquinone glucoside-8-hydroxy-3-methyl-anthraquinone-1-O-(4-O-β-D-galactopyranosyl)- α-L-rhamnopy-ranoside(I), β-sitosterol, stearyl alcohol, quercetin-3-O-β-D-galactopyranoside, leucocyanidin, chrysophanol, emodin and physcion isolated from leaves.

References : Ambasta (1986), Raju (2000), Rastogi and Mehrotra (1993), Singh *et al.* (1996), Watt (1972).

531. *Rumex scutatus* Linn.

Family : Polygonaceae.

English and Hindi Names: **Eng.** : French Sorrel; **Hindi** : Ambavati.

Distribution : W. Himalaya.

Part/s Used : Leaves.

Utilization : Leaves used for flavouring soups and omelettes.

References : Ambasta (1986), Kirtikar and Basu (1984), Mehtani and Ingle (1999).

532. ***Ruta chalepensis*** **Linn.**

Syn : R.angustifolia Pers.; *R.bracteosa* DC.; *R. graveolens* Linn. var. *angustifolia* Hook.f.

Family : Rutaceae.

English, Sanskrit and Hindi Names: **Eng. :** Ave Grace, Common Rue, Countryman's Treacle, Garden Rue, Herb-Grace, Herb of repentance; **Sans. :** Guchhapatra, Pitapushpa, Somalata, Vishapaha; **Hindi :** Pismarum.

Distribution : Cultivated throughout India.

Part/s Used : Leaves.

Utilization : Leaves used as a condiment and for garnishing food products. Also, it yields etheral oil. Fresh herb yields Garden Rue Oil, used for scenting soaps and in perfumery.

Active Principles : Active ingredients are angustifolin, arborinine, bergapten, chalepensin, chalepin, isopimpinellin xanthotoxin, coumarin, heliettin, kokusaginine, skimmianine, rutamarin (aerial parts), 2-undecanone (38.1), 2-nonanyl acetate (20.9), 2-nonanone (17.2), 2-decanyl acetate (5.6%), tridecane-2,4-dione, heptadecane-2,4-dione, α-pinene, sabinene, myrcene, caryophyllene, trans- α-bergamotene, β-elemene, β-cyclocitral, β-phenylethylbutanoate, *cis*-and trans-cinnamoyl isovalerate (essential oil).

References : Ambasta (1986), Bhattacharjee (1998), Chevallier (1996), Kirtikar and Basu (1984), Mehtani and Ingle (1999), Rastogi and Mehrotra (1990, 1991, 1993, 1995), Watt (1972).

533. ***Salix babylonica*** **Linn.**

Family : Salicaceae.

English Name : **Eng. :** The weeping willow.

Distribution : Grown in N. India, both in the plains and Himalaya upto 3,000m.

Part/s Used : Leaves.

Utilization : Young leaves have an odour of roses.

Active Principles : Sucrose and glucose detected in leaves.

References : Ambasta (1986), Rastogi and Mehrotra (1993, 1995), Watt (1972).

534. ***Salix caprea*** **Linn.**

Family : Salicaceae.

English and Hindi Names: **Eng. :** Goat willow, The Sallow; **Hindi:** Bedmushk.

Distribution : Cultivated in N.W. India.

Part/s Used : Flowers. Leaves.

Utilization: Flowers yield an essential oil ('Attar'), used in low grade perfumery. Leaf oil used for making perfumed waters.

Active Principles : Salicin, saligenin, gallocatechin, rutin, cynaroside, quercetin, luteolin, (+) epigallocatechin and (+) gallocatechin dimers, I and II isolated from leaves.

References : Ambasta (1986), Kirtikar and Basu (1984), Rastogi and Mehrotra (1991), Watt (1972).

535. *Salvia officinalis* Linn

Family : Lamiaceae.

English and Hindi Names : **Eng. :** Common sage, Garden sage, Red sage, Sage; **Hindi :** Salvia sefakuss.

Distribution : Native to the Mediterranean, sage is cultivated all round the world, thriving in sunny conditions. Small scale cultivation in Jammu.

Part/s Used : Leaves. Flowers.

Utilization : Dried leaves used as a condiment and used as a flavouring in variety of food preparations. Flowers yield Dalmatian Sage oil, used in perfumery and as a deodorant, and also in place of the leaves for flavouring soups, salads, sandwitches, meats, table sauces, sausages and other canned foods to avoid discoloration caused by leaves.

Active Principles : These are α-and β- thujones, camphor, cineole, α-and β- pinenes, camphene, cymene, limonene, β-myrcene, borneol, linalool and bornyl acetate (essential oil), apigenin-7-O-glucoside, camphor, 1,8-cineole, cirsimaritin, 7-erythrosmanal, luteolin-7-O-glucoside, oleanolic acid, picrosalvin, saccharose, β-sitosterol, ursolic acid and viridifloral (leaves).

References : Ambasta (1986), Chevallier (1996), Chopra *et al.* (1960), Harborne and Baxter (2001), Kirtikar and Basu (1984), Mehtani and Ingle (1999), Raju (2000), Rastogi and Mehrotra (1990, 1991, 1993, 1995), Watt (1972).

536. *Salvia sclarea* Linn.

Family : Lamiaceae.

English Name : **Eng. :** Clary sage.

Distribution : Hilly tracts of H. P. and Uttar Pradesh.

Part/s Used : Flowering Tops. Leaves.

Utilization : Flowers and leaves yield, Oil of Clary Sage, used in high grade perfumery, cosmetics and for flavouring liquors, non-alcoholic and alcoholic beverages, ice-creams, candy and baked goods.

Active Principles : Both the flowers and leaves yield 0.1-0.15% oil, linalyl acetate (48-75%), linalool and other terpene alcohols.

References : Chevallier (1996), Harborne and Baxter (2001), Kumar and Singh (2000), Sharma and Sood (1997).

537. ***Sambucus nigra*** Linn.

Syn : S. vulgaris Lamk.

Family : Caprifoliaceae.

English Name : **Eng. :** European or Black elder.

Distribution : Temperate regions; often cultivated in botanical gardens.

Part/s Used : Flowers.

Utilization : Flowers used mainly for preparation of Elder Flower Water, which is mildly fragrant and used for eye and skin lotions.

Active Principles : These are α- and β- amyrins, campesterol, cholesterol, cycloartenol, cycloeuclenol, lupeol, 24-methylene-cycloartanol, stigmasterol, isorhamnetin-3-O-rutinoside (flowers), palmitic (37.9), linolenic (20.1), linoleic (19.6) and oleic (8.5) acids and alkanes (7.2%) (oil from flowers).

References : Ambasta (1986), Chevallier (1996), Greenish (1999), Rastogi and Mehrotra (1991, 1993), Watt (1972).

538. ***Santalum album*** Linn. (Pl. 10E)

Syn : S. myrtifolium Roxb.; *S. verum* Linn.; *Sirium myrtifolium* Roxb.; *Sandalum album* Rumph.

Family : Santalaceae.

English, Sanskrit and Hindi Names : **Eng. :** Sandal Tree, Sandal Wood, White Sandalwood, Yellow Sandalwood; **Sans. :** Ananditam, Chandana, Taliaparnam; **Hindi :** Chandal, Chandan, Safed-Chandan, Sandal.

Distribution : S.W. India. Karnataka. Tamil Nadu.

Part/s Used : Wood-Oil.

Utilization : Strongly scented Sandalwood oil extracted from the heart-wood widely used in perfumery and forms the base for 'Attars' produced in India. The wood powder is used for making Agarbatti, incense powder, flavouring candies and sherbets.

Active Principles : Trans-β-santalol, epi-cis-β-santalol, α-santalol (52%), β-santalol (23%), β-santalic, teresantalic, nortricyclo-ekasantalic, bicycloekasantalic and dihydro β-santalic acids isolated from sandalwood oil.

References : Ambasta (1986), Bhakuni (2000), Bhattacharjee (1998), Chevallier (1996), Greenish (1999), Jain (1968), Kirtikar and Basu (1984), Kumar and Singh (2000), Mehtani and Ingle (1999), Mittre (1981), Raju (2000), Rastogi and Mehrotra (1991, 1993, 1995), Sahni (1998), Vihari (1995), Watt (1972).

539. *Sapium baccatum* Roxb.

Syn : S. populifolium Wall.; *Carumbium baccatum* Kurz.; *Excoecaria baccata* Muell.; *E. affinis* Grifff ; *Stillingia paniculata* Miq.

Family : Euphorbiaceae.

English Name : **Eng. :** Mousedeer's Rubber Tree.

Distribution : Sikkim Himalaya. Assam. Khasia Hills.

Part/s Used : Fruits.

Utilization : Fruits used as a flavouring.

Active Principles : Taxaxerone, taraxerol, 1-hexacosanol and β-sitosterol isolated from fruits.

References : Ambasta (1986), Rastogi and Mehrotra (1990), Watt (1972).

540. *Sassafras albidium* (Nutt.) Nees

Syn : S. officinale Nees and Eberm.; *S. variifolium* Kuntze; *Laurus sassafras* Linn.

Family : Lauraceae.

English Names: **Eng. :** Ague Tree, Sassafras.

Distribution : Dried roots imported into India.

Part/s Used : Root-bark.

Utilization : Root-bark yields sassafras oil, employed in soaps, cheap perfumes and as a flavouring agent.

Active Principles : Root contains 2-5% of oil, with safrole (80-90%) as the major component along with α-pinene, α- and β-phellandrenes, asrone, 5-methoxy eugenol, apiole, coniferaldehyde, camphor, myristicin, thujone, (-)-menthone, caryophyllene, eugenol, elemicine and copaene. Boldine, cinnamolaurine, norboldine, norcinnamolaurine and reticuline perchlorate isolated from root-bark.

References: Ambasta (1986), Bhattacharjee (1998), Greenish (1999), Harborne and Baxter (2001), Rastogi and Mehrotra (1991), Watt (1972).

541. *Satureja hortensis* Linn.

Family : Lamiaceae.

English Name : **Eng. :** Summer Savory.

Distribution : Kashmir.

Part/s Used : Leaves. Seeds.

Utilization : Leaves used to flavour meat and other culinary preparations. Also, yield an essential oil (oil of Savory), employed for flavouring meat, eggs and vegetables.

Active Principles : Leaves possess apigenin-7-O-glucuronoside, luteolin-7-O-glucoside and luteolin-7-O-glucuronoside.

References : Ambasta (1986), Mehtani and Ingle (1999), Rastogi and Mehrotra (1993).

542. ***Satureja montana*** Linn.

Family : Lamiaceae.

English Name : **Eng.** : Winter Savory.

Distribution : Kashmir.

Part/s Used : Whole Herb.

Utilization : Leaves and Winter Savory oil extracted from dried herb used for flavouring meat and meat products, baked goods, confectionery, condiments and relishes.

Active Principles : 1.6% oil composed mainly of carvacrol, p-cymene and thymol. Plant contains 13.0% tannins, caffeic, chlorogenic and rosmarinic acids.

References : Ambasta (1986), Chevallier (1996), Harborne and Baxter (2001), Rastogi and Mehrotra (1995).

543. ***Saussurea lappa*** C.B. Clarke

Syn : *S.costus* (Falc.) Lipsch.; *Aplotaxis lappa* Dcne.; *Aucklandia costus* Balc.

Family : Lamiaceae.

English, Sanskrit and Hindi Names : **Eng.** : Kuth, The Costus; **Sans.** : Agada, Bhasura, Dushta, Gada, Jarana, Kakala, Kustha, Niruja, Pakala; **Hindi** : Kot, Kur, Kust, Kut, Pachak.

Distribution : W. Himalaya. Kashmir valley: 2,500-4,000m. Kinnaur and Lahoul-Spiti in H.P.

Part/s Used : Roots.

Utilization : Roots used as incense and its essential oil, (Costus oil), used in high class perfumery and cosmetics for blending purposes.

Active Principles : Kuth roots contain resinoids (6%), essential oil (1.5%), alkaloid (0.05%), inulin (18%), a fixed oil and other minor constituents like tannins and sugars. Essential oil contains saussurine.

References : Ambasta (1986), Anonymous (1999), Bhattacharjee (1998), Chevallier (1996), Kirtikar and Basu (1984), Koelz (1979), Mittre (1981), Purohit (2000), Sharma and Sood (1997), Singh (1999), Sood *et al.* (2001), Watt (1972).

544. ***Schinus molle*** Linn.

Family : Anacardiaceae.

English Names : **Eng.** : Mastic Tree or Peruvian Pepper Tree.

Distribution : Planted in Indian gardens and avenues.

Part/s Used : Fruits.

Utilization : Fruits used as a substitute for pepper (*P. nigrum* Linn.). Also, yield an essential oil.

Active Principles : These are β-spathulene (essential oil) and isomasticadienonalic acid (berries).

References : Ambasta (1986), Rastogi and Mehrotra (1991).

545. *Schisandra grandiflora* Hook.f. and Thoms.

Syn : Kadsura grandiflora Wall.; *Sphaerostema grandiflorum* H.f. & T.

Family : Schisandraceae.

Distribution : Temperate Himalaya: 2,000-3,500m. Aka hills in N.E. India.

Part/s Used : Flowers. Seeds.

Utilization : Flowers fragrant. Seeds aromatic.

Active Principles : Schizandraflorin isolated from the plant.

References : Ambasta (1986), Negi and Gaur (1994), Rastogi and Mehrotra (1993), Watt (1972).

546. *Schleichera oleosa* (Lour.) Oken.

Syn : Pistacia oleosa Lour.; *S. trijuga* Willd.

Family : Sapindaceae.

English Name : **Eng. :** Ceylon Oak.

Distribution : Geldubba of Chhindwara district, M.P.

Part/s Used : Seeds.

Utilization : Seed oil employed in perfumery.

Active Principles : Major cyanolipid of oil contains two fatty acids esterified with 1-cyano-2-hydroxymethyl prop-2-en-1-ol.

References : Bennett (1987), Bhattacharjee (1998), Rastogi and Mehrotra (1991).

547. *Scleria biflora* Roxb. Subsp. ***biflora***

Family : Cyperaceae.

Part/s Used : Roots.

Utilization : Roots camphoraceous.

Reference : Ambasta (1986).

548. *Sedum rosea* Scop. *ex* Sprague

Syn : S. imbricatum H.f. & T.; *S. rhodiola* DC.; *Rhodiola imbricata* Edgw.; *R. rosea* Linn.

Family : Crassulaceae.

English Name : **Eng. :** Rose root.

Distribution : W. Alpine Himalaya from Kumaon to Kashmir: 4,000-5,650m.

Part/s Used : Roots.

Utilization : Rootstocks pleasantly odorous with that of roses.

References : Ambasta (1986), Watt (1972).

549. *Selinum tenuifolium* Wall. *ex* Clarke

Syn : S. candolii DC.; *S. tenuifolium* Wall. *ex* Clarke var. *filicifolia* (Edgew.) Clarke.

Family : Apiaceae.

Distribution : Himalaya: 1,800-4,200m.

Part/s Used : Roots.

Utilization : Roots used as a spice.

Active Principles : Bergapten, heraclenin, heraclenol, 8-geranyloxy-psoralen, imperatorin, iso-imperatorin (0.07%), osthol (0.02) and oxypeucedanin (1.2) isolated from roots.

References : Rastogi and Mehrotra (1991, 1993), Sood *et al.* (2001).

550. *Selinum vaginatum* (Edgew.) C.B. Clarke

Family : Apiaceae.

Sanskrit Name : **Sans. :** Bhootkeshi.

Distribution : Himalaya from Kashmir to Kumaon: 1,800-3,900m.

Part/s Used : Roots.

Utilization : Roots aromatic and employed as an incense. Also, used as a cheap substitute for the 'Jatamansi'.

Active Principles : Roots yield anomalin, isopteryxin, selinidin, selinone and vaginatin.

References : Rastogi and Mehrotra (1990, 1991), Sharma and Sood (1997), Singh (1999).

551. *Selinum wallichianum* (DC.) Raizada *et* Saxena

Syn : S. candollii DC.; *S. filicifolium* (Edge.) Nasir.; *S. tenuifolium* Wall *ex* Cl.; *S.t.* var. *filicifolia* (Edge.) Cl.

Family : Apiaceae.

Part/s Used : Fruits. Seeds.

Utilization : Both fruits and seeds used for flavouring foods.

References : Bennett (1987), Negi *et al.* (1993), Negi and Gaur (1994).

552. *Senecio chrysanthemoides* DC.

Syn : S. jacobaea D.Don; *S. sisymbriformis* DC.

Family : Asteraceae.

Distribution : Temperate parts (Kashmir to Sikkim). Khasia mountains.

Utilization : Yields an essential oil, suitable as a perfumery material.

Active Principles : Essential oil contains anhydrooplopanone, germacrene D and 1-undecene.

References : Ambasta (1986), Bennett (1987), Chopra *et al.* (1980), Rastogi and Mehrotra (1990,1993).

553. *Sesamum indicum* Linn. (Pl. 11A)

Syn : S. luteum Retz.; *S. occidentale* Heer and Regel.; *S. orientale* Linn.

Family : Pedaliaceae.

English, Sanskrit and Hindi Names : **Eng. :** Gingelly, Sesame; **Sans. :** Homadhanya, Pavitra, Snehaphala, Tila; **Hindi :** Bariktel, Gingli, Krishna-tel, Mitha-tel, Til ka tel.

Distribution : Cultivated throughout the plains of India.

Part/s Used : Seeds.

Utilization : Seeds used in baked products, sweetmeats and confectionery. Also, yield a fatty oil called Sesame oil, Gingelly oil or Til oil, used as an ingredient of confectionery and in the manufacture of soaps, cosmetics and perfumes.

Active Principles : These are sesamolinol, seasamol, γ-tocopherol (seeds), pinoresinol and sesamolinol (seed oil).

References : Ambasta (1986), Bhattacharjee (1998), Bennett (1987), Chevallier (1996), Greenish (1999), Jain and Sharma (2000), Kirtikar and Basu (1984), Mehtani and Ingle (1999), Rastogi and Mehrotra (1995), Vihari (1995), Watt (1972).

554. *Shorea robusta* Gaertn.f. (Pl. 11B)

Syn : Vatica robusta Steud.

Family : Dipterocarpaceae.

English, Sanskrit and Hindi Names: **Eng. :** Common Sal, Indian Dammer Sal Tree; **Sans. :** Ashvakarna, Chiraparna, Dhanya, Divyasara; **Hindi :** Sakher, Sakhua, Sakoh, Sal, Salwa, Shal.

Distribution : Sub-Himalayan tract from the Kangra valley to Arunachal Pradesh.

Part/s Used : Flowers. Wood.

Utilization : Flowers fragrant. Sal resin used as an incense, and also yields an essential oil, Chua oil, used for flavouring, chewing and smoking tobacco.

Active Principles : Epi-ψ-taraxastanonol, β-sitosterol, hydroxy hopanone, dammarenediol II, dipterocarpol, dammarenolic acid, asiatic acid and α-amyrin isolated from resin.

References : Ambasta (1986), Bhattacharjee (1998), Kirtikar and Basu (1984), Mittre (1981), Rastogi and Mehrotra (1995), Sahni (1998), Watt (1972).

555. *Skimmia arborescens* T. Anders. *ex* Gamble

Syn : S. laureola Hook.f. in part, non Sieb. and Zucc.; *An quetilia laureola* Dcne.; *Limonia laureola* Wall.; *Laureola fragrans* Roem.

Family : Rutaceae.

Distribution : Temperate Himalaya. Khasia Hills.

Part/s Used : Leaves.

Utilization : Leaves scented-like musk and used as a flavouring and incense (`Havan Samagri'). Also, yield a volatile oil, employed in flavouring industry and a base material for high grade perfumes.

References : Ambasta (1986), Mittre (1981), Kirtikar and Basu (1984), Raju (2000), Singh (1999), Watt (1972).

556. *Smilax aristolochiaefolia* Mill.

Family : Liliaceae.

English Name : **Eng. :** Mexican Sarsaparilla.

Part/s Used : Roots.

Utilization : Root extract used as a flavouring ingredient.

Active Principles : Roots mainly constitute sarsapogenin, smilagenin, sitosterol and stigmasterol.

Reference : Bhattacharjee (1998).

557. *Sonneratia caseolaris* (L.) Engl.

Syn : S. acida Linn.f.; *Aubletia caseolaris* Gaertn.; *Rhizophora caseolaris* Linn.

Family : Sonneratiaceae.

Distribution : Tidal creeks and littoral forests of India and Andamans.

Part/s Used : Fruits.

Utilization : Fruits used for flavouring curries and chutneys.

References : Ambasta (1986), Bennett (1987), Kirtikar and Basu (1984), Mehtani and Ingle (1999).

558. *Sorbus aucuparia* Linn.

Syn : Pyrus aucuparia Gaertn.; *P. ursina* Herb, Strach. and Wint.

Family : Rosaceae.

English Names : **Eng. :** Mountain ash, Rowan tree.

Distribution : W. Temperate Himalaya from Kashmir to Kumaon: 3,850-4,350m.

Part/s Used : Ripe Fruits.

Utilization : Ripe fruits used as a flavouring agent in the preparation of extracts, syrups and juices, brandies, liquors, lemonades, marmalades, confectionery, foods and coffee.

Active Principles : Fruits contain asozone, isoquercetin, isoquercitrin, meratin, caffeic, quinic and chlorogenic acids, naringin, (-) epigallocatechin gallate, (-) epigallocatechin gallic acid, phloroglucinol, α-carotene (6.2), neo-β-carotene (3.3), β-carotene (3.8), polycopene (1.4) and ξ-carotane (1.6mg%), ursolic and parascorbic acid.

References : Ambasta (1986), Chevallier (1996), Mehtani and Ingle (1999), Rastogi and Mehrotra (1990, 1991), Watt (1972).

559. *Spondias cytherea* Soon.

Syn : S.dulcis Soland. *ex* Forst. f.

Family : Anacardiaceae.

English Names: **Eng.:** Ambarella, Golden-apple, Great-Hog-Plum, Otaheite apple.

Distribution : Cultivated in India.

Part/s Used : Fruits.

Utilization : Fruits edible, possess flavour-like that of mango or pineapple. Unripe fruits also used to flavour sauces, soups and stews.

References : Ambasta (1986), Watt (1972).

560. *Spondias pinnata* (Linn.f.) Kurz

Syn : S. acuminata Roxb., non Gamble; *S. amara* Lamk.; *S. macrophylla* Wall. *ex* Hook.f.; *S. mangifera* Willd.; *Mangifera pinnata* Linn.f.; *Evia amara* Lamk.

Family : Anacardiaceae.

English, Sanskrit and Hindi Names : **Eng. :** Bile Tree, Indian Hog Plum, Traveller's Delight, Wild Mango; **Sans. :** Amrataka, Bhringiphalla, Kapichuda, Markatamra, Rasadhya; **Hindi :** Amara, Ambodha, Ambra, Jangli am.

Distribution : Sub-Himalayan tracts and outer valleys upto 1,000m.

Part/s Used : Flowers. Fruits.

Utilization : Flowers fragrant and used as a flavouring. Ripe fruits used as a condiment and made into chutneys, pickles, jams, sauces and stews.

Active Principles : Fruits contain β-amyrin, oleanolic acid, alanine, cystine, glycine, serine and leucine.

References : Ambasta (1986), Anonymous (1999), Bajpayee and Dixit (1996), Bennett (1987), Kirtikar and Basu (1984), Mehtani and Ingle (1999), Negi and Gaur (1994), Raju (2000), Rastogi and Mehrotra (1991), Watt (1972).

561. ***Stemodia viscosa*** Roxb.

Syn : S. arvensis Steud.; *S. maritima* Heyne.

Family : Scrophulariaceae.

Distribution : C. and Peninsular India. Throughout the Deccan.

Part/s Used : Whole Plant.

Utilization : Dried plants slightly fragrant.

Active Principles : Diosmetin, its 7-O-β-D-glucuronide and luteolin 7-O-β-D-glucuronide isolated from leaves.

References : Ambasta (1986), Kirtikar and Basu (1984), Rastogi and Mehrotra (1991), Watt (1972).

562. ***Stixis suaveolens*** (Roxb.) Pierre

Syn : Roydsia suaveolens Roxb.

Family : Capparidaceae.

Part/s Used : Fruits.

Utilization : Ripe fruits aromatic.

References : Ambasta (1986), Bennett (1987).

563. ***Strophanthus gratus*** Franch.

Family : Apocynaceae.

English Name : **Eng. :** Smooth Strophanthus.

Distribution : Tropical regions of India; grown as an ornamental plant in gardens.

Part/s Used : Flowers.

Utilization : Flowers fragrant.

References : Bhattacharjee (1998), Rastogi and Mehrotra (1990), Watt (1972).

564. ***Syzygium aromaticum*** (Linn.) Merrill and Perry (Pl. 11C)

Syn : Caryophyllus aromaticus Linn.; *Eugenia aromatica* Kuntze; *E. caryophyllata* Thunb.

Family : Myrtaceae.

English, Sanskrit and Hindi Names: **Eng. :** Clove Tree; **Sans. :** Devapuspa; **Hindi :** Laung.

Distribution : Cultivated in S. India.

Part/s Used : Flower Buds.

Utilization : Dried unopened floral buds, cloves used as a spice and much esteemed to flavour ham, roasts, pickles, preserves, ketchups, sausages, cakes, pies, puddings cookies and candy, wines, chewing tobacco and as an ingredient of betel-chew. Essential oil, called Clove oil, extensively used for flavouring food products, fermented beverages, scenting soaps, toilet waters and in perfumery. Clove stem and leaf oils used as a cheap substitute for clove oil, particularly in frozen meat products.

Active Principles : Clove buds yield 15-20%oil, of which eugenol is 60-90%, eugenol acetate 2-25%, β-caryophyllene 5-12%, farnesol, α-humulene, asarone, vanillin, isoeugenol, and naphthalene.

References : Ambasta (1986), Anonymous (1999), Bhattacharjee (1998), Bennett (19870, Chevallier (1996), Greenish (1999), Harborne and Baxter (2001), Mehtani and Ingle (1999), Rastogi and Mehrotra (1991, 1993, 1995), Watt (1972).

565. ***Syzygium cumini*** (Linn.) Skeels

Syn : S. jambolanum DC; *Eugenia caryophyllifolia* Lam.; *E. cuminii* Druce; *E. jambolana* Lam.

Family : Myrtaceae.

English, Sanskrit and Hindi Names: **Eng. :** Black Plum, Jaman, Jambolan, Jambul Tree, Java Plum; **Sans. :** Brahaspati, Jambavam, Jambu, Jambula, Nilaphala, Rajphala, Shyamala; **Hindi :** Bahojaman, Jaman, Jamun, Jamniphalani, Phalinda.

Distribution : Wild or cultivated over the greater part of India.

Part/s Used : Flowers.

Utilization : Flowers fragrant.

Active Principles : Flowers contain acetyl oleanolic acid, triterpenoids, ellagic acid, isoqercitrin, quercetin, kaemferol and myricetin, myricetin-3-L-arabinoside, dihydromyricetin and quercetin-3-D-galactoside.

References : Ambasta (1986), Anonymous (1999), Bhattacharjee (1998), Bennett (1987), Chevallier (1996), Kirtikar and Basu (1984), Negi and Gaur (1994), Rastogi and Mehrotra (1990, 1991, 1993), Watt (1972).

566. ***Syzygium jambos*** (Linn.) Alston

Syn : Eugenia jambos Linn.; *Jambosa vulgaris* DC.

Family : Myrtaceae.

English, Sanskrit and Hindi Names : **Eng. :** Malay apple, Rose apple; **Sans. :** Brihayphala, Jambu, Kokileshta, Mahaphala, Nanda, Rajajambu, Shukapriya; **Hindi :** Gulab jaman.

Distribution : Cultivated in different parts of India.

Part/s Used : Fruits.

Utilization : Fruits flavour resemble like that of rose water.

Active Principles : These are alanine, glutamine, threonine, tyrosine, aspartic acid and cysteine.

References : Ambasta (1986), Bhattacharjee (1998), Bennett (1987), Kirtikar and Basu (1984), Rastogi and Mehrotra (1990), Watt (1972).

567. *Tagetes lucida* Cav.

Family : Asteraceae.

English Name : **Eng.** : Sweet scented Marigold.

Part/s Used : Leaves. Capitula.

Utilization : Capitula used for scenting bathing water and leaves used in soups.

Reference : Ambasta (1986).

568. *Tagetes minuta* Linn.

Syn : *T. glandulifera* Schrank

Family : Asteraceae.

English Names : **Eng.** : Stinking-Roger, Wild Marigold.

Distribution : Naturalized in the N.W. Himalaya: 1,250-2,500m.

Part/s Used : Capitula.

Utilization : Capitula used as a flavouring ingredient of baked goods, confectionery, frozen dairy desserts, alcoholic and non-alcoholic beverages, condiments, relishes, gelatins, puddings and tobacco. Also, yield Tagetes oil, used as a fragrance component in floral and herbaceous perfumes.

Active Principles : Oil contains ocimene, dihydrotagetone, tagetone, trans-ocimene, β-myrcene, linalool, limonene, α - and β- pinenes, carvone, with lesser amounts of phenylethanol, geraniol, p-cymene, sabinene and cineole.

References : Ambasta (1986), Harborne and Baxter (2001), Rastogi and Mehrotra (1990, 1991, 1995), Singh (1999).

569. *Tagetes patula* Linn.

Family : Asteraceae.

English, Sanskrit and Hindi Names : **Eng.** : French Marigold; **Sans.** : Taugla; **Hindi** : Genda.

Distribution : Cultivated in gardens all over India upto 1,350m.

Part/s Used : Florets. Herb.

Utilization : Florets employed for the production of 'Attare Genda'. Herb also yields a volatile oil used in perfumery.

Active Principles : Allopatuletin isolated from petals.

References : Ambasta (1986), Jain and Sharma (2000), Rastogi and Mehrotra (1993).

570. *Tagetes tenuifolia* Cav.

Syn : T. signata Bartl.

Family : Asteraceae.

English Name: **Eng. :** Striped Marigold.

Distribution : Grown in Indian gardens.

Utilization : Yields a pleasant smelling essential oil.

Active Principles : Dihydrotagetone, tagetone, β-ocimene, (E) 7 (Z) ocimenones extracted from flower essential oil.

References : Ambasta (1986), Rastogi and Mehrotra (1993,1995).

571. *Talauma candolii* Blume

Family : Magnoliaceae.

Distribution : Introduced into Indian gardens.

Part/s Used : Flowers.

Utilization : Flowers fragrant.

Reference : Ambasta (1986).

572. *Tamarindus indica* Linn.

Syn : T. occidentalis Gaertn.; *T. officinalis* Hook.

Family : Caesalpiniaceae.

English, Sanskrit and Hindi Names : **Eng. :** Tamarind Tree; **Sans. :** Amla, Bhukta, Charitra, Sukta, Tintrini; **Hindi :** Amli, Amlika, Anbli, Imli, Nuli, Teter.

Distribution : Most parts of India.

Part/s Used : Fruits.

Utilization : The pulp of fruits has a sweet-sour flavour and included in curries and chutneys. Also, used as a condiment.

Active Principles : Tamarind pulp contains about 10% of free tartaric acid, 8% of acid potassium tartarate, 25-40% of invert sugar and tamarindienal.

References : Ambasta (1986), Bhattacharjee (1998), Chevallier (1996), Greenish (1999), Harborne and Baxter (2001), Jain (1968), Kirtikar and Basu (1984), Mehtani and Ingle (1999), Raju (2000), Rastogi and Mehrotra (1990), Sahni (1998), Singh and Pandey (1998), Watt (1972).

573. *Tanacetum longifolium* Wall. *ex* DC.

Syn. : Chrysanthemum dolichophyllum Kitam.; *T. dolichophyllum* (Kitam.) Kitam.; *T. himachalensis* Aswal and Mehrotra.

Family : Asteraceae.

English Name : **Eng. :** Long Leaved Tansy.

Distribution : Himalaya-Kashmir to Nepal: 3,000-4,200m.

Part/s Used : Roots.

Utilization : Roots used as incense.

Active Principles : Root oil yields terpinen-4-ol, sabinene and p-cymene.

References : Ambasta (1986), Bennett (1987), Rastogi and Mehrotra (1995).

574. *Tanacetum nubigenum* Wall. *ex* DC.

Family : Asteraceae.

Hindi Name : **Hindi :** Dhoop gogul.

Distribution : Himalaya-Ladakh to Arunachal Pradesh: 2,700-3,600m.

Part/s Used : Roots.

Utilization : Both roots and a gum resin ('Gogul') extracted from it used as incense.

Active Principles : Gum resin possesses hydroxyketones-22-hydroxy-octacosan-25-one and 24-hydroxy-triacontane-27-one tetra- and hexacosanoic acids and β-sitosterol.

References : Ambasta (1986), Rastogi and Mehrotra (1995).

575. *Tanacetum tenuifolium* Jacquem. *ex* DC.

Syn : Artemisia tenuiflora Jacq.

Family : Asteraceae.

Distribution : Kumaon.

Utilization : Used for flavouring puddings.

References : Ambasta (1986), Watt (1972).

576. *Tanacetum vulgare* Linn.

Syn : Chrysanthemum vulgare (Linn.) Bernh.

Family : Asteraceae.

English Name : **Eng. :** Tansy.

Distribution : Found as weed in Jammu and Kashmir.

Part/s Used : Flowering Tops. Leaves.

Utilization : Leaves aromatic, used for garnishing and flavouring pudding. Flowering tops yield Tansy oil, used in perfumery.

Active Principles : Oil rich in thujone, camphor, sesquiterpene lactones, flavonoids and resin.

References : Ambasta (1986), Bhattacharjee (1998), Chevallier (1996), Rastogi and Mehrotra (1995).

577. ***Telosma cordata*** (Burm.f.) Merrill

Syn : Pergularia minor Andr.; *T. minor* (Andr.) Craib; *T.odoratissima* Coville.

Family : Asclepiadaceae.

English Names : **Eng. :** Cowslip, E. coast-creeper, Primrose-creeper.

Distribution : Grown in gardens in plains and hills.

Part/s Used : Flowers.

Utilization : Flowers valued in perfumery. Also, yield an essential oil.

Active Principles : Essential oil extracted from flowers contains geraniol, β-ionone, dihydro-β-ionone and cis- and trans- theaspiranes.

References : Ambasta (1986), Rastogi and Mehrotra (1995).

578. ***Tetraclinis articulata*** Mast.

Family : Cupressaceae.

English Name : **Eng. :** Arar-Tree.

Part/s Used : Resin.

Utilization : Resin used for making incense.

Reference : Ambasta (1986).

579. ***Thelepaepale ixiocephala*** (Benth.) Bremek.

Family : Acanthaceae.

Part/s Used : Flowering Tops.

Utilization : Flowering tops and its essential oil camphoraceous.

Reference : Ambasta (1986).

580. ***Theobroma cacao*** Linn.

Family : Sterculiaceae.

English Names : **Eng. :** Cacao, Chocolate Tree, Cocoa.

Distribution : Cultivated in the S. India.

Part/s Used : Seeds (Cocao beans).

Utilization : Powdered cocoa seeds used for flavouring confectionery. Cocoa butter obtained from seeds also used in cosmetics, perfumery, milk chocolates, etc.

Active Principles : These are theobromine (kernels), xanthines (seed pulp) and endorphins (seeds).

References : Ambasta (1986), Bhattacharjee (1998), Chevallier (1996), Greenish (1999), Kirtikar and Basu (1984), Rastogi and Mehrotra (1995), Watt (1972).

581. ***Thlaspi arvense*** Linn.

Family : Brassicaceae.

English Names : **Eng.** : Common Pennycress, Fanweed.

Distribution : Throughout the Temperate and Sub-Alpine Himalaya, ascending to 4,050m.

Part/s Used : Seeds.

Utilization : Seeds eaten with caution as a condiment.

References : Ambasta (1986), Chopra *et al.* (1956).

582. ***Thymus linearis*** Benth.

Syn : *T. serpyllum* auct. (non L.); *T. quinquecostatus* Celak.

Family : Lamiaceae.

Distribution : W. Temperate Himalaya: 1,700-4,400m.

Part/s Used : Flowers. Leaves. Twigs.

Utilization : Dried leaves, flowers and twigs used as condiments to flavour curries.

Active Principles : These are thymol (19.1-64.6), p-cymene (9.1-46.9), carvacrol (1.9-5.0%) (essential oil of leaves and flowers), linalool, carvacrol and geranyl acetate (essential oil of aerial parts). Besides, α-pinene, β-pinene, camphene, car-3-ene, α-terpinene, limonene, γ-terpinene, terpinolene, citronellal and trans-β-terpineol present in essential oil.

References : Rastogi and Mehrotra (1991, 1993, 1995), Sood *et al.* (2001), Watt (1972).

583. ***Thymus serpyllum*** Linn.

Family : Lamiaceae.

English and Hindi Names : **Eng.** : Bank Thyme, Brotherwort, Creeping Thyme, Hill-wort, Horsethyme, Mother of Thyme; **Hindi** : Banajwain.

Distribution : W. Temperate Himalaya: Kashmir to Kumaon.

Part/s Used : Flowering Tops. Leaves.

Utilization : Flowering tops added for flavouring cordial drinks. Leaves and flowering tops yield a volatile oil, Wild Thyme, employed as a cheap flavouring agent.

Active Principles : Oil from fresh leaves and flowering tops contains scutellarin heteroside, luteolin-7-glucoside, apigenin, thymol, p-cymene and carvacrol.

References : Ambasta (1986), Bhattacharjee (1998), Chevallier (1996), Kaul (1997), Kirtikar and Basu (1984), Mehtani and Ingle (1999), Gupta (1981), Purohit (2000), Rastogi and Mehrotra (1995).

584. *Thymus vulgaris* Linn.

Family : Lamiaceae.

English Name : **Eng. :** Common Thyme, Garden Thyme.

Distribution : Grown in N. India upto 900m.

Part/s Used : Leaves. Flowers.

Utilization : Powdered dried leaves and flowering tops constitute the fragrant Thyme, used for garnishing and flavouring various meat and fish curries. Thyme oil (extracted from its leaves and flowers) used for flavouring meat, sausages, sauces, canned foods, and also in cheap perfumery.

Active Principles : Thyme oil contains thymol, γ-terpinene, p-cymene, (E)-2-hexenol, bornyl acetate, carvacrol, hexane, (E)-2-hexenal, myrcene, linalool, β-caryophyllene, 1-octen-3-ol, α-thujene, trans-sabinene hydrate and α-terpinene.

References : Ambasta (1986), Bhattacharjee (1998), Chevallier (1996), Chopra *et al.* (1960), Mehtani and Ingle (1999), Rastogi and Mehrotra (1990, 1991, 1993, 1995), Raju (2000).

585. *Tigridia pavonia* Ker-Gawl.

Family : Iridaceae.

English Names: **Eng. :** Shell Flower, Tiger Flower.

Part/s Used : Bulbs.

Utilization : Bulbs flavour - like chestnut.

Reference : Ambasta (1986).

586. *Toddalia asiatica* (L.) Lam.

Syn : Limonia oligandra Dals.; *Paullinia asiatica* Linn.; *Scopolia aculeata* Sm.; *Toddalia aculeata* Pers.; *T. angustifolia* Lamk.; *T. angustifolia* Miq.; *T. floribunda* Wall.; *T. nitida* Lamk.; *T. rubricaulis* Willd.; *Zanthoxylon floribundum* Wall.; *Z. nitidum* Wall.

Family : Rutaceae.

English, Sanskrit and Hindi Names : **Eng. :** Forest Pepper, Lopez-root Tree; **Sans. :** Dahana, Kanchana; **Hindi :** Dahan, Jangli Kalimirch, Kanj.

Distribution : Deccan. All districts of Chennai. Khasia Hills.

Part/s Used : Bark. Leaves.

Utilization : Leaves highly aromatic and considered a good substitute for curry leaves (*Murraya koenigii*). Bark yields an essential oil.

Active Principles : Myrcene, (+) limonene, geraniol, citronellol, geranyl acetate, ocimene, car-3-ene, bisabolene, linalool, citronellal and limettin extracted from essential oil.

References : Bennett (1987), Kirtikar and Basu (1984), Nazarudeen *et al.* (1996), Rastogi and Mehrotra (1993), Watt (1972).

587. *Tricalysia sphaerocarpa* (Hook.f.) Gamble

Syn : Diplospora sphaerocarpa Hook.f.

Family : Rubiaceae.

Distribution : W. Peninsula from Bombay southwards.

Part/s Used : Seeds.

Utilization : Roasted seeds taste and smell like coffee and used as its substitute.

Active Principles : Alkaloid like caffeine isolated.

References : Ambasta (1986), Chopra *et al.* (1956).

588. *Trigonella corniculata* Linn.

Syn : T. fimbriata Royle

Family : Papilionaceae.

Sanskrit and Hindi Names : **Sans. :** Malya; **Hindi :** Kasuri methi, Kasturi methi, Marwari methi, Champa methi.

Distribution : Bengal. Kumaon to Kashmir: 1,700-4,000m.

Part/s Used : Plant.

Utilization : Grown as a pot herb for flavouring.

Active Principles : These are diosgenin, tigogenin gitogenin, (leaves and flowers) and ethyl-α-D-galactopyranoside (leaves).

References : Ambasta (1986), Kirtikar and Basu (1984), Rastogi and Mehrotra (1990, 1991).

589. *Trigonella foenum-graecum* Linn. (Pl. 12A)

Family : Papilionaceae.

English, Sanskrit and Hindi Names : **Eng. :** Fenugreek, Greek Hayes; **Sans. :** Asumodhagam, Chandrika, Methika; **Hindi :** Methi, Muthi.

Distribution : Cultivated in many parts of India; wild in N. India.

Part/s Used : Leaves. Seeds.

Utilization : Dried leaves and fenugreek seeds used as a condiment and spice, and also used for flavouring pickles and food preparations. They are a component of curry powders, bread doughs, cheese, chutneys and various spice blends. Fenugreek essence finds applicaton in the preparation of an imitation maple syrup and as a seasoning for meat, coffee, vanilla extracts, chutney tobacco and cheese.

Active Principles : Chief components are β-sitosterol, kaempferol, quercetin, saponins-graecunins A, B, C, D, E, F and G (leaves), tigogenin, neotigogenin, diosgenin, yamogenin, gitogenin, 4-hydroxy isoleucine, vitexin, isovitexin, saponaretin, homoorientin, vicenin-1 and vicenin-2 (seeds).

References : Ambasta (1986), Anonymous (1999), Bhattacharjee (1998), Chevallier (1996), Greenish (1999), Kirtikar and Basu (1984), Lakshamanan *et al.* (1986), Mehtani and Ingle (1999), Shah (1984), Singh and Pandey (1998), Watt (1972).

590. *Trilisa odoratissima* Cass.

Family : Asteraceae.

English Names : **Eng. :** Carolina Vanilla, Vanilla Plant.

Part/s Used : Leaves.

Utilization : Leaves smell of vanilla and used for flavouring tobacco.

Active Principles : Leaves rich in coumarin.

Reference : Bhattacharjee (1998).

591. *Triphasia trifolia* (Burm.f.) P.Wils.

Syn : Limonia trifoliata Linn.; *L. diacantha* DC.; *T. aurantiola* Lour.; *T. trifoliata* DC.

Family : Rutaceae.

English and Hindi Names : **Eng. :** Chinese Lime, Limeberry, Myrtle-Lime; **Hindi :** Cheeninaranghi.

Distribution : Escape in the W. Peninsula and in gardens throughout other regions of India.

Part/s Used : Fruits.

Utilization : Fruits aromatic; and employed in jellies, marmalades and pickles.

Active Principles : Semi-β-carotenone and β-carotenone isolated.

References : Ambasta (1986), Bennett (1987), Rastogi and Mehrotra (1990, 1995), Watt (1972).

592. *Tsuga dumosa* (D.Don) Eichler

Syn : T. brunoniana Carr.

Family : Pinaceae.

English Names : **Eng. :** Himalayan Hemlock Spruce, Indian Hemlock Fir.

Distribution : E. Kumaon. Bhutan to Arunachal Pradesh: 2,400-3,000m.

Utilization : Various parts of plant used as incense.

References : Ambasta (1986), Sahni (1998).

593. *Usnea longissima* Ach.

Family : Usneaceae.

Distribution : Arunachal Pradesh, H.P., W. Bengal.

Part/s Used : Lichen Thallus.

Utilization : Used as spices and for flavouring non-vegetarian preparations, pulses and other vegetables Also, used as 'dhoop' (incense).

Active Principles : Thallus contains usnic, barbatic, barbatolic, diffractic, squamatic, fumarprotocetraric and evernic acids.

References : Mudgal and Hajra (1997), Upreti (1996).

594. ***Usnea lucea***

Family : Usneaceae.

Distribution : Tamil Nadu.

Part/s Used : Lichen.

Utilization : As spices.

Active Principles : Stictic acid.

Reference : Mudgal and Hajra (1997).

595. ***Valeriana hardwickii*** Wall.

Syn : V. acuminata Royle; *V. elata* Don.; *V. javanica* Blume; *V. tenera* Wall.

Family : Valerianaceae.

Hindi Names: **Hindi :** Asarun, Shumeo, Tagger.

Distribution : Temperate Himalaya (Kashmir to Bhutan): 1,350-4,000m. Khasia mountains: 1,350-2,000m.

Part/s Used : Whole Plant.

Utilization : Whole plant including rhizome and roots employed as an incense. Roots used mainly in perfumery to scent and clean the hair (Irvine).

References : Ambasta (1986), Kirtikar and Basu (1984), Saklani and Jain (1994), Watt (1972).

596. ***Valeriana jatamansi*** Jones

Syn : V. spica Vahl; *V. villosa* Wall.; *V. wallichii* DC.

Family : Valerianaceae.

English, Sanskrit and Hindi Names: **Eng. :** Indian Velerian; **Sans. :** Barhana, Chakra, Danda, Hasti, Kutil, Nata; **Hindi :** Mushkbala.

Distribution : Temperate Himalaya at 3,500m. Khasia and Jaintia Hills.

Part/s Used : Rhizomes. Roots.

Utilization : Rhizomes and roots used as incense and a fixative for blending in high grade perfumes. They also yield essential oil, Valeriana oil, used as an adjunct to certain flavours for tobacco, honey, root-beer, etc.

Active Principles : These are isovaleric and caproic acids (rhizomes and roots), heptadecanyl eicosanoate and methyl eicosanoate (rhizomes), α-pinene, limonene, 1,8-cineole, p-cymene, borneol acetate, borneol, nerolidol and maaliol sesquifenchene (root oil).

References : Ambasta (1986), Bennett (1987), Anonymous (1999), Kaul (1997), Kirtikar and Basu (1984), Mehtani and Ingle (1999), Misra and Kumar (2000), Purohit (2000), Sharma and Sood (1997), Singh (1999), Rastogi and Mehrotra (1990, 1991, 1993), Watt (1972).

597. *Valeriana officinalis* Linn.

Syn : *V. dubia* Bunge

Family : Valerianaceae.

English Names : **Eng.** : Capon's Tail, Cat's Valerian, Common Valerian, Great Wild Valerian, Great Valerian Healall, Setwall.

Distribution: Temperate Himalaya upto 3,000m. Khasia Hills. Cultivated in Kashmir, H.P. and Uttar Pradesh.

Part/s Used : Rhizomes. Roots. Flowers.

Utilization : Rhizomes and roots yield an essential oil, Valerian Oil used for flavouring in tobacco, food products and beverages and in perfumery. Flowers also fragrant.

Active Principles : Fresh roots and rhizome extracts contain cardiovalen. Yield of oil is 0.5-2%, and comprehends (-)-bornyl acetate, bornyl isovalerate, caryophyllene α- and pinenes, valerenal, valerenone, valerenic acid, isovaleric acid, eugenyl isovalerate, isoeugenyl isovalerate, β-ionone, valerianol, patchouli alcohol, borneol, camphene, β-bisabolene and terpinolene.

References : Ambasta (1986), Bhattacharjee (1998), Chevallier (1996), Chopra *et al.* (1980), Greenish (1999), Harborne and Baxter (2001), Kirtikar and Basu (1984), Mehtani and Ingle (1999), Kumar and Dwivedi (2000), Sinha and Sinha (2001), Watt (1972).

598. *Vallaris glabra* Kuntze

Syn : *V. pergulanus* Burm.f.

Family : Apocynaceae.

Part/s Used : Flowers.

Utilization : Flowers fragrant; also contain essential oil.

Reference : Ambasta (1986).

599. *Vanilla planifolia* Andr.

Syn : *V. fragrans* Ames.

Family : Orchidaceae.

English Name : **Eng.** : Vanilla.

Distribution : Cultivated in Kerela, Tamil Nadu, Karnataka, Lakshadweep, Andaman and Nicobar Islands and certain parts of N.E. India.

Part/s Used : Pods.

Utilization : Flowers sweet-smelling. Pods, costliest spice crop, constitute the vanilla of commerce, and employed for flavouring chocolates, beverages, confections, cakes, custards, puddings and ice-creams, and also in the manufacture of soaps, perfumes and sachet powders.

Active Principles : Vanillin is the chief odorous constituent of vanilla pods. Other components are p-hydroxy benzaldehyde and p-hydroxy benzyl methyl ether.

References : Ambasta (1986), Bajpayee and Dixit (1996), Bhatta-charjee (1998), Greenish (1999), Mehtani and Ingle (1999), Watt (1972).

600. *Vanugeria tomentosa* Hochst.

Family : Rubiaceae.

Part/s Used : Fruits.

Utilization : Fruits have a good flavour - like medlars.

Reference : Vickery (1990).

601. *Vatica lancaeifolia* Blume

Syn : *V. canaca* Ham.; *Vateria lanceaefolia* Roxb.; *V. lanceolaria* Roxb.; *V. lanceolata* Roxb.

Family : Dipterocarpaceae.

English Name : **Eng.** : Mascal – Wood.

Distribution : E. Himalaya. Assam. E. Bengal.

Part/s Used : Bark exudates.

Utilization : Tree yields an oleoresin used for flavouring tobacco.

References : Ambasta (1986), Mehtani and Ingle (1999), Watt (1972).

602. *Vetiveria zizanioides* (Linn.) Nash

Syn : *Anatherum muricatum* Retz.; *A. zizanioides* (Linn.) Hitchcock and Chase; *Andropogon muricatus* Retz.; *A. squarrosus* Hook.f., non Linn. f.; *Phalaris zizanoides* Linn.; *Rhaphis muricata* Nees.; *V. odorata* Virey.

Family : Poaceae.

English, Sanskrit and Hindi Names : **Eng.** : Cuscus, Khas-Khas, Khus-Khus, Koosa, Vetiver; **Sans.**: Abhaya, Haripriya, Reshira, Sugandhimula; **Hindi** : Bala, Balah, Bena, Khas-Khas, Panni.

Distribution : Throughout India.

Part/s Used : Roots.

Utilization : Fragrant roots yield an essential oil, Vetiver oil, much valued in high grade perfumery, cosmetics and for scenting `sherbets'.

Active Principles : These are khusinol, isokhusinol oxide, methoxyphenol, o-, m- and p- cresols, eugenol, 4-vinyl guaiacol, cis- and trans-isoeugenol, 4-vinyl phenol, vanillin, isovalencenol, zizanol, p-vetinone and zizanoic acid (Vetiver oil), tricyclic sesquiterpenoids-zizanal, epizizanal together with α- and β- vetvones, khusimone and bicyclodecenone (roots).

References : Ambasta (1986), Harborne and Baxter (2001), Kirtikar and Basu (1984), Kumar and Singh (2000), Rastogi and Mehrotra (1990, 1991, 1995), Watt (1972).

603. *Viburnum foetens* Dcne.

Family : Caprifoliaceae.

Distribution : N.W. Himalaya: 1,700-3,650m.

Part/s Used : Flowers.

Utilization : Flowers fragrant sweetly odorous.

References : Negi and Gaur (1994), Watt (1972).

604. *Viburnum grandiflorum* Wall. *ex* DC.

Syn : V.nervosum Hook.f. and Thoms.

Family : Caprifoliaceae.

Hindi Names: **Hindi :** Telanu, Thekla.

Distribution : Himalaya (Kashmir to Sikkim): 3,650-4,350m.

Part/s Used : Flowers.

Utilization : Flowers find use in perfumery.

References : Ambasta (1986), Bennett (1987), Rastogi and Mehrotra (1991), Watt (1972).

605. *Viburnum mullaha* Ham.

Syn : V. stellulatum Wall.

Family : Caprifoliaceae.

Distribution : N.W. Himalaya: 2,000-3,650m.

Part/s Used : Fruits.

Utilization : Dried fruits used as flavouring.

References : Negi and Gaur (1994), Watt (1972).

606. *Vigna vexillata* (Benth.) A.Rich.

Family : Papilionaceae.

Part/s Used : Rootstock.

Utilization : Tuberous rootstock considered superior to sweet-potato in flavour and nutrients.

Reference : Ambasta (1986).

607. ***Viola odorata*** Linn.

Family : Violaceae.

English, Sanskrit and Hindi Names : **Eng.** : Apple-leaf, Bairnwort, Banwort Bessy Banwood, Blaver, Blue Violet, English Violet, Fine-leaf, March Violet, Sweet Violet, Vilip; **Sans.** : Nilapushpa, Sukshmapatra, Vanapsa; **Hindi** : Banafshah.

Distribution : N.W. Himalaya: 1,700-2,000m.

Part/s Used : Leaves.

Utilization : Leaf absolute used in high grade perfumery.

Active Principles : Dodecan-1-ol, pentadeca-5,10-dien-1-ol, pentadec-5-en-1-ol, octadec-1-ene, eicos-1-ene and octadeca-9,12,dienoic acid isolated from leaves.

References : Ambasta (1986), Chevallier (1996), Kirtikar and Basu (1984), Rastogi and Mehrotra (1995), Watt (1972).

608. ***Vitis parvifolia*** Roxb

Syn : *V. truncata* Miq.; *V. vulpina* Linn.

Family : Vitaceae.

English Name: **Eng.** : Himalyan Wild Vine.

Distribution : Sub-Temperate valleys of the N.W. Himalaya.

Part/s Used : Flowers. Fruits.

Utilization : Flowers and fruits pleasantly odorous.

References : Ambasta (1986), Negi and Gaur (1994), Watt (1972).

609. ***Ximenia americana*** Linn.

Syn : *X. russelliana* Wall.; *X. spinosa* Salisb.

Family : Olacaceae.

English Names: **Eng.** : False Sandalwood, Mountain Plum, Tallow Wood.

Distribution : E. and W. Peninsulas. Andamans.

Part/s Used : Wood.

Utilization : Wood pleasantly scented and used in place of sandal wood.

References : Ambasta (1986), Kirtikar and Basu (1984), Mehtani and Ingle (1999), Watt (1972).

610.*Xylopia parvifolia* Hook.f. and Thoms.

Family : Annonaceae.

English Names : **Eng. :** Atuketiya, Netawu.

Part/s Used : Flowers. Fruits. Root-bark.

Utilization : Flowers, fruits and root-bark sweet scented and chewed with betel leaves.

Reference : Ambasta (1986).

611.*Zanthoxylum acanthopodium* DC.

Syn : Z. alatum Roxb. and its var. *planispinum* (Sieb. and Zucc.) Rehd. and E.H. Wils.; *Z. alatum* Wall.; *Z. armatum* DC.; *Z. hostile* Wall.; *Z. planispinum* Sieb. And Zucc.

Family : Rutaceae.

Sanskrit and Hindi Names : **Sans.:** Dhiva, Gandhalu, Tejohva, Tamburu, Sauraja; **Hindi :** Darmar, Nepalidhaniya, Tejphal, Timur, Tumra.

Distribution : Hot valleys of the Sub-Tropical Himalaya upto 2,500m. Khasia Hills, 1,750-2,000m.

Part/s Used : Flowers. Fruits. Seeds.

Utilization : Young twigs, roasted seeds and fruits used as spice and condiments. Aromatic seeds also find their use for making perfumed tobacco and Henna 'Attra'. Fruits have a coriander-like flavour and used as spice. Dried flowers and fruits yield an oil of Wartara, used as deodorant and in perfumery.

Active Principles: These are linalool, α-pinene, cumene, limonene, phellandrene, 2-heptanol, p-cymene, linalool, decylaldehyde, geranyl acetate, neryl acetate, geraniol, α-selenene and citral (flower essential oil), linalool, limonene, methyl cinnamate and cineole (fruits), linalool (50.72-64.9%) (fruit oil), tambulin (3,5-dihydroxy-7,8,4'-trimethoxyflavone) and tambulol (glucosyl-7,4' dimethyl ether of gossypetin) (fruits), 6-hydroxy nonadec-4 (Z)-enoic, 8-hydroxy pentadec-4 (Z) enoic, 7-hydroxy-7-vinylhexadec-4 (Z) enoic acids (seeds), linalool (58.3), limonene (24.46) and methyl cinnamate (8.92%) (seed essential oil).

References : Ambasta (1986), Anonymous (1999), Bennett (1987), Bhattacharjee (1998), Chandra (1997), Kirtikar and Basu (1984), Kulkarni and Kumbhojkar (1993), Lal *et al.* (1996), Misra and Kumar (2000), Negi and Gaur (1994), Rastogi and Mehrotra (1991, 1993), Saklani and Jain (1994), Singh (1999), Singh and Singh (1985), Sinha and Sinha (2001), Watt (1972).

612. *Zanthoxylum limonella* (Dennst.) Alston

Syn : Fagara budrunga Roxb.; *F. rhetsa* Roxb.; *Tipalia limonella* Dennst., Schluss.; *Z. budrunga* Wall. *ex* DC.; *Z. crenatum* Wall.; *Z. rhetsa* (Roxb.) DC.

Family : Rutaceae.

Sanskrit and Hindi Names: **Sans. :** Ashvaghra, Atitejani, Bidalaghni, Parijata, Sutejasi, Tejasvini, **Hindi :** Badrang, Jaladhari, Pepuli.

Distribution : Tropical Himalaya. Khasia Hills. E. Bengal.

Part/s Used : Leaves. Seeds.

Utilization : Tender leaves used as a condiment. Seeds have a black pepper - like taste.

References : Ambasta (1986), Bennett (1987), Kirtikar and Basu (1984), Rastogi and Mehrotra (1995), Watt (1972).

613. *Zanthoxylum rhetsa* (Roxb.) DC.

Syn : Fagara budrunga Roxb., not of Wall.; *F. oblongum* Wall.

Family : Rutaceae.

Distribution : W. Peninsula.

Part/s Used : Fruits.

Utilization : Fruits used as condiments to flavour curries.

Active Principles : Fruits rich in zanthoxylin, arborine and dictamnine.

References : Mittre (1981), Rastogi and Mehrotra (1993, 1995), Saklani and Jain (1994), Watt (1972).

614. *Zanthoxylum nitidum* (Roxb.) DC.

Syn : Fagara nitida Roxb.; *Z. hamiltonianum* Wall. *ex* Hook.f.; *Z. h.* var. *tomentosum* Hook.f.

Family : Rutaceae.

Distribution : Sikkim. Assam.

Part/s Used : Fruits. Seeds.

Utilization : Aromatic fruits used as condiments. Seed oil resembles that of bergamot and geranium oils.

Active Principles : These are β-phellandrene, phenylethyl acetate, methyl cinnamate, (essential oil from seeds), linalool and β-phellandrene (seeds).

References : Ambasta (1986), Bennett (1987), Kirtikar and Basu (1984), Rastogi and Mehrotra (1991), Watt (1972).

615. *Zanthoxylum ovalifolium* Wight

Syn : Limonia leptostachya Jack.; *Loddalia mitis* Miq.; *Z. lucidum* Wall.; *Z. ovalifolium* Wight var. *sepiarium* (Wight) Hook.f.

Family : Rutaceae.

Distribution : Kanara. Nilgiri Hills. Khasia Hills. Assam.

Part/s Used : Seeds.

Utilization : Seed husk rich in fragrant essential oil.

References : Ambasta (1986), Bennett (1987), Watt (1972).

616.*Zanthoxylum oxyphyllum* Edgew.

Syn : Z. violaceum Wall.

Family : Rutaceae.

Distribution : Throughout the Hills of Kumaon: 2,000-3,000m. Bhutan. Khasia hills.

Part/s Used : Fruits.

Utilization : Fruits used as a spice and condiment for flavouring curries.

Active Principles : Sesamin, eudesmin, epieudesmin, syringare-sinol, γ-fagarine, β-sitosterol and lupeol isolated from fruits.

References : Ambasta (1986), Dam and Hajra (1981), Kirtikar and Basu (1984), Saklani and Jain (1994), Watt (1972).

617.*Zingiber cassumunar* Roxb.

Syn : Z. cliffordii Andhr.; *Z. montanum* auct. Non (Koenig) Dietr.: Ramamoorthy; *Z. purpureum* Roscoe.

Family : Zingiberaceae.

Sanskrit and Hindi Names : **Sans. :** Vanardraka; **Hindi :** Banada.

Distribution : Throughout India.

Part/s Used : Rhizomes.

Utilization : Rhizomes added to food preparations for flavour. Yields an essential oil, used in perfumery for the preparation of artificial geranium, pepper and rose oil for scenting soaps.

Active Principles : These are α- and β-pinenes, d-camphene, zingiberene, curcumene, camphor, linalool, α-terpineol, d-borneol, zingiberol (rhizome essential oil), cis-3-(2',4'-5'-trimethoxyphenyl)-4-[(E)-2',4',5'-trimethoxystyrnyl] cyclohex-1-ene (I) and 8-(3',4'-dimethoxyphenyl)-2-methoxynaphtho-1,4-quinone (II)-(rhizomes).

References : Ambasta (1986), Bennett (1987), Kirtikar and Basu (1984), Mehtani and Ingle (1999), Rastogi and Mehrotra (1991), Watt (1972).

618.*Zingiber chrysanthemum* Rosc

Family : Zingiberaceae.

Distribution : Uttar Pradesh. Sikkim. Assam ascending to 1,600m.

Part/s Used : Fruits. Rhizomes.

Utilization : Fruits aromatic. Rhizomes yield an essential oil.

Active Principles : Rhizome essential oil rich in 1,8-cineole (42.0), α-fenchyl acetate (15.0), camphene (6.0) and α-terpneol (6.0%).

References : Ambasta (1986) Rastogi and Mehrotra (1995).

619. ***Zingiber elatum*** Roxb.

Family : Zingiberaceae.

Part/s Used : Rhizomes.

Utilization : Rhizomes camphoraceous, used in perfumery.

Reference : Ambasta (1986).

620. ***Zingiber montanum*** (Koen.) Thel.

Family : Zingiberaceae.

Part/s Used : Rhizome.

Utilization : Rhizome used as spices and condiments for flavouring food preparations and as a substitute for true ginger.

References : Mittre (1981), Tripathi *et al.* (2001).

621. ***Zingiber officinale*** Rosc. (Pl. 12B, 12C)

Syn : Amomum zingiber Linn.

Family : Zingiberaceae.

English, Sanskrit and Hindi Names : **Eng. :** East Indian Ginger, Ginger, Ginger Root; **Sans. :** Anupama, Ardraka, Chandrakhya, Vara; **Hindi :** Ada, Adrak.

Distribution : Cultivated all over India.

Part/s Used : Rhizome.

Utilization : Dried ginger widely used as a spice and condiment for flavouring pies, puddings, cookies, pastries, biscuits, etc. whereas green ones is used mainly in culinary preparations, pickles, canned ginger and ginger cocktails. Also yields an essential oil, oil of ginger or the oleoresin, for flavouring confectionery items, soft drinks (mainly gingerale), table sauces, pickles, canned ginger, and also in perfumery.

Active Principles : Major constituents of oil are zingiberene, neral, geranial, ar-curcumene, 1,8-cineole, p-cymene, camphene, zerumbone, nerolidol, terpinen-4-ol, propionaldehyde, linalool, α- and β-pinenes, nerol, citronellol, α-terpineol, 1-nonanol, limonene, β-phellandrene, β and γ bis abolenes, β-sesquiphellandrene, 6-,8-, and 10-gingerols.

References : Ambasta (1986), Anonymous (1999), Chevallier (1996), Greenish (1999), Katz (1996), Kirikar and Basu (1984), Mehtani and Ingle (1999), Rastogi and Mehrotra (1991, 1995), Tripathi *et al.* (2001), Watt (1972).

622. ***Zingiber zerumbet*** Rosc. *ex* Sm.

Syn : Amomum zerumbet Linn.; *A. spurium* Gmel.; *Z. spurium* Kon.

Family : Zingiberaceae.

Sanskrit and Hindi Names : **Sans. :** Ahava, Avanti, Karpooraharidra, Viranam; **Hindi :** Mahabaribach, Narkachur.

Distribution : Throughout India.

Part/s Used : Rhizome.

Utilization : Rhizome used as a condiment and spice and its oil used for scenting soaps and in toiletries.

Active Principles : Essential oil contains zerumbone (8-oxohumulene), humulene monoxide and humulene dioxide, two ketones (~6), a diepoxide (~4), two 2-α-epoxides-humulene epoxide I and humulene epoxide II (65) and humulenol (~3) alongwith (+) α-curcumene (2%) and caryophyllene epoxide.

References : Ambasta (1986), Arora (1981), Islam (1996), Kirtikar and Basu (1984), Rastogi and Mehrotra (1990, 1993, 1995), Watt (1972).

623. ***Zizyphus mauritiana*** Lamk.

Syn : Rhamnus jujuba Linn.; *Z. jujuba* Lamk. (non Mill.); *Z. scororia* Schult.

Family : Rhamnaceae.

English, Sanskrit and Hindi Names : **Eng. :** Chinese Date, The Indian Jujube; **Sans. :** Ajapriya; **Hindi :** Pewandi.

Distribution : Throughout drier regions of India and ascending to 1,400m. in the Himalaya.

Part/s Used : Fruits.

Utilization : Ripe fruits pleasantly odorous.

Active Principles : Fruits rich in p-coumaroylates of maslinic acid-3-O-trans-p-coumaroyl- and 3-O-cis-p-coumaroyl maslinic acids together with betulonic, oleanonic and maslinic acids as their methyl esters.

References : Banerjee and Ghora (1996), Bennett (1987), Raju (2000), Rastogi and Mehrotra (1991, 1993, 1995), Sahni (1998), Watt (1972).

6

CONCLUSION

Information on six hundred and twenty five species of flavouring and fragrant plants of India has been recorded in this documentation. They are represented by 121 families and 331 genera (Appendix I). These include mostly the angiospermous taxa (580 spp in 356 genera), followed by lichens [(23 spp in 11 genera: *Anaptychia* (1), *Evernia* (1), *Everniastrum* (2), *Heterodermia* (1), *Lobaria* (1), *Parmelia* (5), *Parmotrema* (3), *Ramalina* (4), *Rimelia* (1), *Roccella* (2), *Usnea* (2)], gymnosperms [(16 spp in 9 genera: *Cedrus* (1), *Chamaecyparis* (1), *Cryptomeria* (1), *Cupressus* (2), *Juniperus* (6), *Picea* (1), *Pinus* (2), *Tetraclinis* (1), *Tsuga* (1)], algae [(3 spp in 3 genera: *Hydrodictyon* (1), *Laurentia* (1), *Palmaria* (1)] and pteridophytes [(3 spp in 2 genera: *Phymatodes* (2), *Pteris* (1)]. It is apparent that the dicotyledonous taxa comprising 468 spp under 298 genera predominate in their use.

It is evident from Annexure II that the families with the largest number of spp employed for flavouring and fragrance purposes by the ethnic communities of India are: Lamiaceae (56 spp), followed by Apiaceae (35 spp), Zingiberaceae (35 spp), Asteraceae (31 spp), Rutaceae (28 spp), Rosaceae (27 spp), Liliaceae (19 spp), Lauraceae (17 spp), Myrtaceae (14 spp), Poaceae (13 spp), Brassicaceae (12 spp), Papilionaceae(12 spp), Oleaceae (11 spp), Parmeliaceae (11 spp), Piperaceae (11 spp), Anacardiaceae (10 spp), Rubiaceae (10 spp), Annonaceae (9 spp), Cupressaceae (9 spp), Clusiaceae (8 spp), Cyperaceae (8 spp), Apocynaceae (7 spp), Burseraceae (7 spp), Pinaceae (7 spp), Usneaceae (7 spp), Arecaceae (6 spp), Magnoliaceae (6 spp), Mimosaceae (6 spp), Solanaceae (6 spp), Araceae (5 spp), Boraginaceae (5 spp), Ericaceae (5 spp), Euphorbiaceae (5 spp), Malvaceae (5 spp), Moraceae (5 spp), Polygonaceae (5 spp), Ranunculaceae (5 spp), Verbenaceae (5 spp), Asclepiadaceae (4 spp), Capparidaceae (4 spp), Caprifoliaceae (4 spp), Cucurbitaceae (4 spp), Geraniaceae (4 spp), Sapotaceae (4 spp), Scrophulariaceae (4 spp), Valerianaceae (4 spp), Caesalpiniaceae (3 spp), Dipterocarpaceae (3 spp), Elaeagnaceae (3 spp), Iridaceae (3 spp),

Thymelaeaceae (3 spp). Of the remaining 70 families, Altingiaceae, Bombacaceae, Caryophyllaceae, Chenopodiaceae, Convolvulaceae, Dilleniaceae, Dipsacaceae, Erythroxylaceae, Meliaceae, Myristicaceae, Olacaceae, Orchidaceae, Oxalidaceae, Pandanaceae, Passifloraceae, Physciaceae, Polypodiaceae, Phytolaccaceae, Rhamnaceae, Roccellaceae, Salicaceae, Sapindaceae, Sonneratiaceae and Urticaceae are represented by 2 spp each whereas the remaining 46 families are represented by 1 sp each.

It is imperative from the present compilation that the genera, viz., *Allium* (12 spp), *Cinnamomum* (10 spp), *Jasminum* (9 spp), *Mentha* (9 spp), *Piper* (9 spp), *Artemisia* (8 spp), *Cymbopogon* (8 spp), *Rosa* (8 spp), *Citrus* (7 spp), *Prunus* (7 spp), *Amomum* (6 spp), *Curcuma* (6 spp), *Cyperus* (6 spp), *Juniperus* (6 spp), *Ocimum* (6 spp), *Zanthoxylum* (6 spp), *Zingiber* (6 spp), *Alpinia* (5 spp), *Carum* (5 spp), *Garcinia* (5 spp), *Lavendula* (5 spp), *Parmelia* (5 spp), *Acacia* (4 spp), *Annona* (4 spp), *Commiphora* (4 spp), *Leucas* (4 spp), *Pelargonium* (4 spp), *Ramalina* (4 spp), *Tagetes* (4 spp), *Tanacetum* (4 spp), *Abelmoschus* (3 spp), *Blumea* (3 spp), *Bupleurum* (3 spp), *Eucalyptus* (3 spp), *Hedychium* (3 spp), *Kaempferia* (3 spp), *Michelia* (3 spp), *Nepeta* (3 spp), *Nigella* (3 spp), *Parmotrema* (3 spp), *Pistacia* (3 spp), *Rhododendron* (3 spp), *Rubus* (3 spp), *Selinum* (3 spp), *Syzygium* (3 spp), *Thymus* (3 spp), *Valeriana* (3 spp) and *Viburnum* (3 spp) are significant in uses (Annexure III). Some of these spp seem to be either less known or unknown and hold considerable scope for popularisation and commercial exploitation by improving upon their cultivation practices. It is hoped that this invaluable data would be of immense use to the flavour industry in the country.

7

LITERATURE CITED

Abraham, Z. 1981. Ethnobotany of the Todas, the Kotas and the Irulas of the Nilgiris : 308-320. In Jain, S.K. (*ed.*) *Glimpses of Indian Ethnobotany*. Oxford and IBH Publishing Co., New Delhi.

Abraham, Z. 1990. Further ethnobotanical study of the Todas and Kotas of the Nilgiris : 255-260. In Jain, S.K. (*ed.*) *Contribution to Indian Ethnobotany*. Scientific Publishers (India), Jodhpur.

Agrawal, S.R. 1981. Trees, flowers and fruits in Indian folk songs, folk proverbs and folk tales: 3-12. In Jain, S.K. (*ed.*) *Glimpses of Indian Ethnobotany*. Oxford and IBH Publishing Co., New Delhi.

Ambasta, S.P. (*ed.*) 1986. *The Useful Plants of India*.CSIR, New Delhi.

Anonymous. 1999. *The Ayurvedic Pharmacopoeia of India*. Part I. Vol II. Government of India- Ministry of Health and Family Welfare. Department of ISM & H.

Arctander, S. 1960. *Perfume and Flavour Materials of Natural Origin*. Elizabeth, New Jersey, U.S.A.

Arora, R.K. 1981. Native food plants of the Northeastern tribals : 91-106. In Jain, S.K. (*ed.*) *Glimpses of Indian Ethnobotany*. Oxford and IBH Publishing Co., New Delhi.

Asolkar, L.V., Kakkar, K.K. and Chakre, O.J. 1992. *Second Supplement to Glossary of Indian Medicinal Plants with Active Principles*. Part-I (A-K), CSIR, New Delhi.

Atal, C.K. and Kapur, B.M. 1982. *Cultivation and Utilization of Aromatic Plants*. RRL, CSIR, Jammu Tawi.

Badwhar, R.L., Rao, P.S. and Sethi, H. 1964. *Some Useful Aromatic Plants*. Forest Research Institute, Dehra Dun.

Bajpayee, K.K. and Dixit, G. 1996. Ethnobotanical studies on food-stuffs of tribals of Tarai region, Uttar Pradesh : 128-132. In Maheshwari, J.K. (*ed.*) *Ethnobotany in South- Asia*. Scientific Publishers, Jodhpur, India.

Banerjee, R.N. and Ghora C. 1996. On the domestic use of some unreported plants of west Dinajpur district (West Bengal) : 325-328. In Maheshwari, J.K. (*ed.*) *Ethnobotany in South- Asia*. Scientific Publishers, Jodhpur, India.

Bennett, S.S.R. 1987. Name changes in Flowering Plants of India and Adjacent Regions. Triseas Publishers, Dehradun, India.

Bhakuni, D.S. 2000. Medicinal plants of commerce of Uttarkhand problems of phyto-pharmaceutical industries : 35-43. In Nautiyal A.R., Nautiyal M.C. and Purohit A.N. (*ed.*) *Harvesting Herbs-2000. Medicinal and Aromatic Plants. An Action Plan for Uttarkhand.* Bishen Singh Mahendra Pal Singh, Dehradun.

Bhattacharjee, S.K. 1998. *Handbook of Medicinal Plants.* Pointer Publishers, New Delhi.

Borathakur, S.K. 1981. Studies in ethnobotany of the Karbis (Mikirs); plant masticatories and dyestuffs : 181-190. In Jain, S.K. (*ed.*) *Glimpses of Indian Ethnobotany.* Oxford and IBH Publishing Co., New Delhi.

Chadha, Y.R. 1982. Flavours and their Industrial Applications (PAFAI, Bombay).

Chandra, Veena 1997. *Edible Plants of Forestry Origin.* ICFRE. Dehradun.

Charalambous, G. and Inglett, G.E. 1978. *Flavour of Foods and Beverages; Chemistry and Technology.* Academic Press, New York.

Chaudhuri, R.H.N. and Pal, D.C. 1981. Plants in folk religion and mythology : 59-68. In Jain, S.K. (*ed.*) *Glimpses of Indian Ethnobotany.* Oxford and IBH Publishing Co., New Delhi.

Chaudhuri, R.H.N., Banerjee D.K. and Guha, A. 1977. Ethnobotanical uses of herbaria. *Bull. Bot. Surv. India.* 19 : 256-261.

Chevallier, Andrew 1996. *The Encyclopedia of Medicinal Plants.* Dorling Kindersley.

Chopra R.N., Chopra I.C. and Varma B.S. 1980. *Supplement to Glossary of Indian Medicinal Plants.* Publications and Information Directorate, New Delhi.

Chopra, I.C., Abrol, B.K. and Handa, K.L. 1960. *Medicinal Plants of the Arid Zones.* United Nations Educational, Scientific and Cultural Organization.

Chopra, R.N., Nayar, S.L. and Chopra, I.C. 1956. *Glossary of Indian Medicinal Plants.* CSIR, New Delhi.

Dam, D.P. and Hajra, P.K. 1981. Observations on ethnobotany of the Monpas of Kameng district, Arunachal Pradesh : 107-114. In Jain, S.K. (*ed.*) *Glimpses of Indian Ethnobotany.* Oxford and IBH Publishing Co., New Delhi.

Datar, R. and Vartak, V.D. 1975. Enumeration of wild edible plants from Karnala bird sanctuary, Maharashtra state. *Biovigyanam* 1 :123-129.

Devon, T.K. and Scott, A.I. 1975. *Handbook of Naturally Occurring Compounds.* Vols I and II. Academic Press Inc., New York.

Duhoon, S.S., Koppar, M.N. and Chandra Umesh 1996. Seabuck thorn (*Hippophae* spp.)- A less known wonder plant of ethno-medico-botanical importance in cold desert of India : 43-45. In Maheshwari, J.K. (*ed.*) *Ethnobotany in South-Asia.* Scientific Publishers (India), Jodhpur, India.

Francis, P. 1984. Plants as human adornment in India. *Econ. Bot.* 38 :194-209.

Frederic, Rosengarten Jr. 1969. *The Book of Species.* Livingston Publishing Co., Wynnewood, Pennsylvania , U.S.A.

Genders, R. 1977. *Scented flora of the World*: 560. Robert Hale Ltd., London. St. Martin's Press, New York.

Girach, R.D. 2001. Ethnobotanical notes on some plants of Mahendragiri hills, Orissa. *Ethnobotany* 13 : 80-83.

Greenish, Henry G. 1999. *Materia Medica*. Scientific Publishers, Jodhpur (India).

Guenther, E. 1948-52. *The Essential Oils.* Vol. I - VI. D. Van. Nostrand Co., New York

Gupta, Rajendra 1981. Plants in folk medicine of the Himalayas: 83-90. In Jain, S.K. (*ed.*) *Glimpses of Indian Ethnobotany*. Oxford and IBH Publishing Co., New Delhi.

Hajra, P.K. and Baishya, A.K. 1981. Ethnobotanical notes on the Miris (Mishinga) of Assam plains : 161-169. In Jain, S.K. (*ed.*) *Glimpses of Indian Ethnobotany*. Oxford and IBH Publishing Co., New Delhi.

Hajra, P.K. and Chakraborty, P.C. 1981. A survey of wild plants in the Lal Market of Gangtok. *Indian. J. Forestry* 4 : 217-220.

Harborne, Jeffrey B. and Baxter, Herbert 2001. *Chemical Dictionery of Economic Plants*. John Wiley and Sons.

Heath, H.B. 1978. *Flavor Technology: Profiles, Products, Applications.* The AVI Publishing Co. Inc., Westport.

Heath, H.B. 1981. *Source Book of Flavours.* The AVI Publishing Co., Inc., Westport.

Hegnauer, R. 1982. Distribution and chemotaxonomy of essentials: 1-23 In (Atal, C.K. and Kapur, B.M *eds.*) *Cultivation and Utilization of Aromatic Plants.* Regional Research Laboratory, CSIR, Jammu-Tawi.

Hussain, Akhtar, Virmani, O.P., Sharma, Ashok, Kumar Anup and Misra, L.N. 1988. Major essential oil-bearing plants of India. CIMAP, Lucknow. In Jain, S.K. (*ed.*) *Glimpses of Indian Ethnobotany*. Oxford and IBH Publishing Co., New Delhi.

Islam, M. 1996. Ethnobotany of certain underground parts of north-eastern region, India : 338-343. In Maheshwari, J.K. (*ed.*) *Ethnobotany in South – Asia*.Scientific Publishers, Jodhpur, India.

Jain, S.K. 1964. Wild food plants of the tribals of Bastar (Madhya Pradesh). *Proc. Nat. Inst. Sci. India* 30B : 56-80.

Jain, S.K. 1967. Ethnobotany – Its scope and study. *Indian Mus. Bull.* 2: 39-43.

Jain, S.K. 1991. *Dictionery of Indian Folk Medicine and Ethnobotany*. Deep Publications, New Delhi.

Jain, S.K. and De, J.N. 1966. Observations on ethnobotany of purulia district, West Bengal. *Bull. Bot. Surv. India*: 8 : 237-251.

Jain, S.K. and Sharma, P.P. 2000. A review of the parallel or unique indigenous uses of twenty five plants in India and some African countries. *Ethnobotany* 12:51-55.

Jain, S.K. 1968. *Medicinal Plants*. National Book Trust, India.

Joshi, P. 1982. An ethnobotanical studies of Bhils-A preliminary survey. *J. Econ. Taxon. Bot.* 3 : 257-266.

Kapur, S.K. 1996a. Traditionally important medicinal plants of Bhaderwah hills – Jammu province. III : 62-69. In Maheshwari, J.K. (*ed.*) *Ethnobotany in South-Asia*. Scientific Publishers (India), Jodhpur, India.

Kapur, S.K. 1996b. Traditionally important medicinal plants of Bhaderwah hills – Jammu province. IV . 70-74. In Maheshwari, J.K. (*ed.*) *Ethnobotany in South-Asia*. Scientific Publishers (India), Jodhpur, India.

Kapur, S.K. and Nanda, S. 1996. Traditionally important medicinal plants of Bhaderwah hills – (Jammu province) II : 56-61. In Maheshwari, J.K. (*ed.*) *Ethnobotany in South- Asia*. Scientific Publishers (India), Jodhpur, India.

Kapur, S.K., Nanda, S. and Srivastava, T.N. 1996a. Ethnobotanical uses of RRL – Herbarium – III : 51-55. In Maheshwari, J.K. (*ed.*) *Ethnobotany in South- Asia*. Scientific Publishers (India), Jodhpur, India.

Kapur, S.K., Nanda, S. and Srivastava, T.N. 1996b. Traditionally important medicinal plants of Bhaderwah hills – Jammu province. III: 62-69. In Maheshwari, J.K. (*ed.*) *Ethnobotany in South- Asia*. Scientific Publishers (India) Jodhpur, India.

Katz, Esther 1996. Emergency foods of the Mixtec highlands (Mexico) : 54-61. In Jain, S.K. (*ed.*) *Ethnobiology in Human Welfare*. Deep Publications, New Delhi.

Kaul, M.K. 1997. *Medicinal Plants of Kashmir and Ladakh*. Indus Publishing Company, New Delhi.

Kirtikar, K.R. and Basu, B.D. 1984. *Indian Medicinal Plants*. Vols. I- IV. Lalit Mohan Basu, Allahabad.

Koelz, Walter H. 1979. Notes on the ethnobotany of Lahul, a province of the Punjab. *Quart. J. Crude Drug. Res.* 17 : 1-56.

Kshirsagar, S.R., Parabia, M.H. and Reddy, M.N. 2003. Ethnobotany of coastal areas in south Gujarat. *Ethnobotany* 15 : 60-63.

Kulkarni, D.K. and Kumbhojkar, M.S. 1993. Kitchen garden plants of Mahadeokoli tribe in Maharashtra. *Ethnobotany* 5 : 119-127.

Kumar, S. and Dwivedi, S. 2000.Agrotechnological gaps and focussed required interventions for commercial cultivation of various medicinal and aromatic plants : 111-124. In Nautiyal A.R., Nautiyal M.C. and Purohit A.N. (*ed.*) *Harvesting Herbs. Medicinal And Aromatic Plants. An Action Plan For Uttarkhand.* Bishen Singh Mahendra Pal Singh, Dehradun.

Kumar, S. and Singh, A.K. 2000. Research and development in medicinal and aromatic plants : 77-97. In Nautiyal A.R., Nautiyal M.C. and Purohit A.N. (*ed.*): *Harvesting Herbs.* Medicinal and Aromatic Plants. An Action Plan For Uttarkhand. Bishen Singh Mahendra Pal Singh, Dehradun.

Lakshmanan, K.K., Subramanian, M.S., Kalyani, K. and Sankaranarayanan, A.S. 1986. Native uses of plants in the Nilgiris. *Proc. Nat. Acad. Sci. India* 56(B): 258-265.

Lal, B., Vats, S.K., Singh, R.D. and Gupta, A.K. 1996. Plants used as ethnomedicine and supplementery food by Gaddis of Himachal Pradesh, India. In Jain, S.K. (*ed.*) *Ethnobiology in Human Welfare*. Deep Publications, New Delhi, : 384-387 pp.

Lalramnghinglova, H. 2000. Ethnobotanical study on the edible plants of Mizoram. *Ethnobotany* 14 : 23-33.

Lalramnghinglova, J.H. 1996. Ethnobotany of Mizoram-a preliminary survey : 439-459. In Maheshwari, J.K. (*ed.*) *Ethnobotany in South Asia.* Scientific Publishers (India), Jodhpur, India.

Lawless, J. 1995. The Illustrated Encyclopedia of Essential Oils. Element, Shaftesbury UK.

Loewenfeld, C. and Back., P. 1974. *The Complete Book of Herbs and Spices.* David and Charles, Melliton Abbot, London.

Maheshwari, J.K. (*ed.*) 1996. *Ethnobotany in South-Asia*. Scientific Publishers (India), Jodhpur, India.

Mandal, S.K. and Basu, S.K. 1996. Ethnobotanical studies among some tribals of Nilgiri district, Tamil Nadu : 268-271. In Maheshwari, J.K. (*ed.*) *Ethnobotany in South- Asia*. Scientific Publishers (India), Jodhpur, India.

Mehra, K.L., Kanodia, K.C. and Srivastava, R.N. 1975. Folk uses of plants for adornment in India, *Econ. Bot.* 29 : 39-46.

Mehtani, S. and Ingle, P.K. 1999. *Plant Food Flavours*. National Institute of Science Communication, New Delhi.

Merory, J. 1968. *Food Flavourings-Compositions, Manufacture and Use*. 2nd Ed. The AVI Publishing Co., Inc., Westport.

Misra, L.N. and Kumar, S. 2000. Indian medicinal plants as the potential source of therapeutic agents : 13-27. In Nautiyal A.R., Nautiyal M.C. and Purohit A.N. (*ed.*) *Harvesting Herbs*-2000. *Medicinal And Aromatic Plants. An Action Plan For Uttarkhand*. Bishen Singh Mahendra Pal Singh, Dehradun.

Mittre, V. 1981. Wild plants in Indian folk life-a historical perspective : 37-58. In Jain, S.K. (*ed.*) *Glimpses of Indian Ethnobotany*. Oxford and IBH, New Delhi.

Mudgal, V. and Hajra, P.K. 1997. *Floristic Diversity and Conservation Strategies in India* . Vol. I. Botanical Survey of India, Calcutta.

Nazarudeen, A., Seeni, S., Koshy, K.C. and Pushpangadan, P. 1996. Folk plants of food, medicine, adornment and repellant used by the Paniyar community in North Kerela : 299-305. In Maheshwari, J.K. (*ed.*) *Ethnobotany in South- Asia*. Scientific Publishers (India), Jodhpur, India.

Negi, K.S. and Gaur, R.D. 1994. A contribution to the edible wild fruits of Uttar Pradesh Hills. *Bull. Bot. Surv. India* 33 : 235-266.

Negi, K.S. and Pant, K.C. 1990. Notes on ethnobotany of the Gangwal – a tribe of Garhwal Himalaya. *Ethnobotany* 2 : 81-85.

Negi, K.S., Gaur, R.D. and Tiwari, J.K. 1999. Ethnobotanical notes on the flora of Har-Ki-Doon (District Uttarkashi), Garhwal Himalaya, Uttar Pradesh, India. *Ethnobotany* 11 : 9-17.

Negi, K.S., Tiwari, J.K., Gaur, R.D. and Pant, K.C. 1993. Notes on ethnobotany of five districts of Garhwal Himalaya, Uttar Pradesh, India. *Ethnobotany* 5 : 73-81.

Painuli, R.M. and Maheshwari, J.K. 1996. Some interesting ethnomedicinal plants used by Sahariya tribe of Madhya Pradesh : 179-185. In Maheshwari, J.K. (*ed.*) *Ethnobotany in South- Asia*. Scientific Publishers (India), Jodhpur, India.

Patole, S. N. and Jain, A.K. 2002. Some wild edible plants of Pachmarhi biosphere reserve (M.P). *Ethnobotany* 14 : 48-51.

Poucher, W.A. 1974. *Perfumes, Cosmetics and Soaps*. Vols. I-III. Chapman and Hall, London.

Prasad N.P. and Abraham Z. 1984. Ethnobotany of Nayadis of North Kerela. *J. Econ. Tax. Bot.* 5: 41-48.

Prasad, P.N., Singh, A.J., Ranjit, A., Narayanan, L.M. and Natarajan, C.R. 1996. Ethnobotany of the Kanikkars of South Tamil Nadu-I : 292-298. In Maheshwari,

J.K. (*ed.*) *Ethnobotany in South - Asia*. Scientific Publishers (India), Jodhpur, India.

Pruthi, J.S. 1976. *Spices and Condiments*. National Book Trust, India.

Pruthi, J.S. 1987. *Bibliography on 57 Spices and Condiments.* (1974-86) pp. 1-168. (over 1600 references). Assoc. Food Scientists and Technologists (India) Delhi.

Pruthi, J.S. 1991a. *Major Spices of India - Crop Management and Post-Harvest Technology*. 1st *Ed.* Pp. 1-650. Indian Council of Agri. Research, New Delhi.

Pruthi, J.S. 1991b. *Minor Spices and Condiments - Crop Management and Post-Harvest Technology*. 1st *Ed.* Pp. 1-650. Indian Council of Agri. Research, New Delhi.

Pruthi, J.S. 1991c. *Modern Methods of Analysis of Spices and Spice Products.* Ist *Ed.* Pp. 1-350. Springer-Verlag, Berlin, Germany, FRG.

Pruthi, J.S. 1991d. *Advances (Innovations) in Spice Processing Technology.* Presented at the 3rd meeting of *International Spices Group* (ISG), Jamaica. pp. 1-25.

Pruthi, J.S. 1992. *Post Harvest Technology of Spices and Condiments.* J. Spices and Aromatic Crops, pp. 1-29. National Research Centre for Spices, Calicut, Kerala.

Purohit, A.N. 2000. Medicinal plants. Need for upgrading technology for trading the traditions : 49-75. In Nautiyal A.R., Nautiyal M.C. and Purohit A.N. (*ed.*): *Harvesting Herbs*-2000. *Medicinal And Aromatic Plants. An Action Plan For Uttarkhand.* Bishen Singh Mahendra Pal Singh, Dehradun.

Rajan, Baskar G., Chezhiyan N., Khan, Irfan Ali and Atiya Khanum. 2003. Spices Medicine: 115-173. In Khan, Irfan Ali and Atiya Khanum (*ed.*) *Role of Biotechnology In Medicinal and Aromatic Plants.* Vols. VIII. Ukaaz Publications, Hyderabad-36.

Raju, R.A. 2000. *Wild Plants of Indian Sub-Continent and Their Economic Use.* CBS Publishers and Distributors, New Delhi.

Ram, Mohan H.Y. and Ram, Manasi 1989. *Fragrant Plants*: 231-266. In Swaminathan, M.S. and Kochhar, S.L. (*ed.*) *Plants and Society.* McMillan Publishers.

Ramachandran, V.S. and Nair, V.J. 1981. Ethnobotanical studies in Cannanore district, Kerela state (India). *J. Econ. Taxon. Bot.* 2 : 65 –72.

Rao, N. R. and Henry, A.N. 1996. *The Ethnobotany of Eastern Ghats in Andhra Pradesh, India.* Botanical Survey of India.Calcutta.

Rao, V.M.K. and Shanpru, R. 1981. Some plants in the life of the Garos of Meghalaya : 153-160. In Jain, S.K. (*ed.*) *Glimpses of Indian Ethnobotany*. Oxford and IBH Publishing Co., New Delhi.

Rastogi, R.P. and Mehrotra, B.N. 1990. *Compendium of Indian Medicinal Plants.* Vol. I. 1960-1969. CDRI, Lucknow and Publications and Information Directorate, New Delhi.

Rastogi, R.P. and Mehrotra, B.N. 1991. *Compendium of Indian Medicinal Plants.* Vol. II. 1970-1979. CDRI, Lucknow and Publications and Information Directorate, New Delhi.

Rastogi, R.P. and Mehrotra, B.N. 1993. *Compendium of Indian Medicinal Plants.* Vol. III. 1980-1984. CDRI, Lucknow and Publications and Information Directorate, New Delhi.

Rastogi, R.P. and Mehrotra, B.N. 1995. *Compendium of Indian Medicinal Plants.* Vol. IV. 1985-1989. CDRI, Lucknow and Publications and Information Directorate, New Delhi.

Rastogi, R.P. and Mehrotra, B.N. 1995. *Compendium of Indian Medicinal Plants.* Vol. V. 1990-1994. CDRI, Lucknow and Publications and Information Directorate, New Delhi.

Rawat, G.S. and Pangtey, Y.P.S. 1987. A contribution to the ethnobotany of alpine regions of Kumaon. *J. Econ. Taxon. Bot.* 11 : 139-148.

Rehman, M.A. and Yusuf, M. 1996. Diversity, ecology and ethnobotany of the Zingiberaceae of Bangladesh : 13-19. In Maheshwari, J.K. (*ed.*) *Ethnobotany in South- Asia.* Scientific Publishers (India), Jodhpur.

Rogers, J.A. 1981. *Essential Oils, in Encyclopaedia of Chemical Technology*:307-332. Vol. XVI.John Wiley and Sons, Inc., New York.

Rosemary, Hemphill 1966. *The Penguin Book of Herbs and Spices.* Penguin Books Ltd., Harmondsworth, Middlesex, England.

Sahni, K.C. 1998. *The Book of Indian Trees.* The Bombay Natural History Society and

Saini, D.C. 1996. Ethnobotany of Tharus of Basti district, Uttar Pradesh : 138-153. In Maheshwari, J.K. (*ed.*) *Ethnobotany in South Asia.* Scientific Publishers (India), Jodhpur.

Saklani, A. and Jain, S.K. 1994. *Cross Cultural Ethnobotany of Northeast India.* Deep Publications, New Delhi.

Sarin, Y.K. 1989. Practical application of ethnobotany. Economic development of backward people : 167-178. In Jain, S.K. (*ed.*) *Methods and Approaches in ethnobotany.* Soc. Ethnobot., CDRI, Lucknow.

Saxena, H.O. and Dutta, P.K. 1975. Studies on the ethnobotany of Orissa. *Bull. Bot. Surv. India* 17 :124-131.

Seth, S.K., Raizada, M.B. and Khan, M.A.W. 1962. *Trees for Vana Mahotsava.* The Manager of Publications, Delhi.

Shah, G.L. 1984. Some economically important plants of Salsette Island near Bombay. *J. Econ. Taxon. Bot.* 5 : 753-765.

Shah, N.C. 2002. Some experience in the field of Ethnobotany. *Ethnobotany* 14: 63-72.

Sharma, K.R. and Sood, M. 1997. *Important Medicinal Plants of Himachal Pradesh.* Vol. I. Directorate of Extension Education. Dr. Y.S. Parmar University of Horticulture and Forestry, Nauni – Solan (H.P.).

Shiva, M.P., Lehri, A., Shiva, A. 2002. *Aromatic and Medicinal Plants.* Intl. Book Distrib. Dehra Dun, India.

Singh, B., Chaurasia, O.P. and Jadhav, K.L. 1996. An ethnobotanical study of Indus Valley (Ladakh): 92-101. In Maheshwari, J.K. (*ed.*) *Ethnobotany in South Asia.* Scientific Publishers (India), Jodhpur.

Singh, H.B. and Visvanathan, M.V. 1996. Useful pteridophytes of India – a gift of nature to human beings : 24-36. In Maheshwari, J.K. (*ed.*) *Ethnobotany in South Asia.* Scientific Publishers (India), Jodhpur.

Singh, P.B. 1999. *Commercially Important Medicinal And Aromatic Plants of Himachal Pradesh.* Society for Herbal medicine and Himalayan Biodiversity, Mandi, H.P.

Singh, S.R. and Singh, N.I. 1985. A preliminary ethnobotanical study on wild edible plants in the markets of Manipur – I. *J. Econ. Taxon. Bot.* 6 : 699-703.

Singh, V. and Pandey, R.P. 1998. *Ethnobotany of Rajasthan, India.* Scientific Publishers (India), Jodhpur.

Sinha, R.K. and Sinha, S. 2001. Medicobotany and biochemistry of some medicinal plants used by the traditional healers of India. Their traditional and modern uses : 127-176. (*ed.*). In *Ethnobiology.* Surabhi Publications. Jaipur.

Smit, H.F. 2000. The universal interest of traditional medicine : 1-11. In Nautiyal A.R., Nautiyal M.C. and Purohit A.N. (*ed.*): *Harvesting Herbs*-2000. *Medicinal and Aromatic Plants. An Action Plan For Uttarkhand.* Bishen Singh Mahendra Pal Singh, Dehradun.

Sood, S.K., Nath Ram and Kalia, D.C. 2001. *Ethnobotany of Cold Desert Tribes of Lahoul-Spiti (N.W. Himalaya).* Deep Publications, New Delhi.

Srivastava, T.N. and Shyama, M. 1963. An ethnobotanical study of Tali (*Dalbergia sissoo* Roxb.) *J. Res. Indian Med. Yoga and Homeo.* 1 : 101-103.

Tarafder, C.R. 1983. On two less known and unknown edible plants used by the tribals in Hazaribagh district of Bihar. *Folklore* 24 : 88-89.

Tripathi, S. and Goel, A.K. 2001. Ethnobotanical diversity of Zingiberaceae in north-eastern India. *Ethnobotany* 13 : 67-79.

Uniyal, B.P. and Malhotra, C.L. 1989. Economic exploitation of rare north-western Himalaya Plants. In Paliwal, G.S. (*ed.*) *The Vegetational Wealth of the Himalayas.* Puja Publishers, Delhi.

Upadhyay, R. and Chauhan, S.V.S. 2000. Ethnobotanical observations on Koya tribe of Gundala mandal of Khamman district, Andhra Pradesh. *Ethnobotany* 12:93-99.

Upreti, D.K. 1996. Studies in Indian ethnolichenology – an overview : 413-414. In Jain, S.K. (*ed.*) *Ethnobiology in Human Welfare.* Deep Publications, New Delhi.

Upreti, D.K., Pant, V. and Divakar, P.K. 2002. Exploitation of lichens from Pithoragarh district, Uttaranchal. *Ethnobotany* 14 : 60-62.

Verrill, A.H. 1940. *Perfumes and Spices.* L.C. Page and Co. Boston.

Vickery, A.R. 1990. Ethnobotanical notes in the herbarium of the Natural history museum, London: I. Rubiaceae. *Ethnobotany* 2 : 25-30.

Vihari, V. 1995. Ethnobotany of Cosmetics of Indo-Nepal border. *Ethnobotany* 7:89-94.

Watt, G. 1889-1899. *A Dictionary of the Economic Products of India.* Vol. 1-6. Periodicals Experts, New Delhi.

Winton, A.L. and Winton, K.B. 1939. *Structure and Composition of Foods.* Vol. IV, John Wiley & Sons, Inc., New York.

APPENDICES

Total Number of Genera and Species of Flavouring and Fragrant Plants under Various Divisions of Plant Kingdom

	Total No. of Genera and Species	Generic Name and Total number of Species Used for Flavouring and Fragrance.
ALGAE	3/3	*Hydrodictyon* (1), *Laurentia* (1), *Palmaria* (1).
LICHENS	11/23	*Anaptychia* (1), *Evernia* (1), *Everniastrum* (2), *Heterodermia* (1), *Lobaria* (1), *Parmelia* (5), *Parmotrema* (3), *Ramalina* (4), *Rimelia* (1), *Roccella* (2), *Usnea* (2).
PTERIDOPHYTES	2/3	*Phymatodes* (2), *Pteris* (1).
GYMNOSPERMS	9/16	*Cedrus* (1), *Chamaecyparis* (1), *Cryptomeria* (1), *Cupressus* (2), *Juniperus* (6), *Picea* (1), *Pinus* (2), *Tetraclinis* (1), *Tsuga* (1).
ANGIOSPERMS		
MONOCOTS	58/112	*Acorus* (1), *Aframomum* (1), *Allium* (12), *Alpinia* (5), *Amomum* (6), *Amphilophis* (1), *Anadendrum* (1), *Ananas* (1), *Antidesma* (1), *Anthoxanthum* (1), *Areca* (1), *Arecastrum* (1), *Chenopodium* (2), *Cocos* (1), *Crocus* (1), *Curcuma* (6), *Cymbopogon* (8), *Cyperus* (6), *Debregeasia* (1), *Dioscorea* (1), *Elettaria* (1), *Etlingera* (2), *Eragrostiella* (1), *Gastrochilus* (1), *Globba* (1), *Hedychium* (3), *Hemerocallis* (1), *Homalomena* (1), *Hyphaena* (1), *Iris* (1), *Jatropha* (1), *Joannesia* (1), *Kaempferia* (3), *Lilium* (2), *Monstera* (1), *Narcissus* (2), *Oncosperma* (1), *Oryza* (1), *Ottelia* (1), *Pandanus* (2), *Phyllanthus* (1), *Pilea* (1), *Polianthes* (1), *Polygonum* (2), *Raphia* (1), *Remirea* (1), *Remusatia* (1), *Rheum* (1), *Rhynchostylis* (1), Rumex (2), *Salix* (2), *Santalum* (1), *Sapium* (1), *Scleria* (1), *Smilax* (1), *Tigridia* (1), *Vanilla* (1), *Vetiveria* (1), *Zingiber* (6).

DICOTS	298/468	*Abelmoschus* (3), *Acacia* (4), *Achillea* (1), *Achras* (1), *Acronychia* (1), *Adansonia* (1), *Aegle* (1), *Aeolanthus* (1), *Aganosma* (1), *Ageratum* (1), *Aisandra* (1), *Alliaria* (1), *Aloysia* (1), *Altingia* (1), *Amaranthus* (1), *Ancistrocladus* (1), *Anethum* (2), *Angelica* (2), *Anisomeles* (1), *Annona* (4), *Anthemis* (1), *Anthocephalus* (1), *Apium* (1), *Aquilaria* (1), *Ardisia* (1), *Armoracia* (1), *Artabotrys* (1), *Artemisia* (8), *Artocarpus* (1), *Azadirachta* (1), *Bauhinia* (1), *Begonia* (1), *Betula* (1), *Blighia* (1), *Blumea* (3), *Borago* (1), *Boswellia* (1), *Bouea* (1), *Brassica* (2), *Bunium* (1), *Bupleurum* (3), *Bursera* (1), *Butea* (1), *Calamintha* (1), *Callicarpa* (1), *Callistemon* (1), *Calotropis* (1), *Camellia* (1), *Cananga* (1), *Canarium* (1), *Cannabis* (1), *Capparis* (1), *Capsicum* (2), *Cardamine* (1), *Careya* (1), *Carissa* (1), *Carum* (5), *Caryocar* (1), *Caryophyllus* (1), *Cassia* (1), *Cedrela* (1), *Cestrum* (1), *Chlorophora* (1), *Chrysanthemum* (2), *Cinnamomum* (10), *Citrus* (7), *Clausena* (2), *Clematis* (1), *Cleome* (1), *Coffea* (2), *Coleus* (1), *Combretum* (1), *Commiphora* (4), *Cordia* (1), *Coriandrum* (1), *Crypteronia* (1), *Cucumis* (1), *Cuminum* (1), *Cydonia* (1), *Daphne* (1), *Daucus* (2), *Decalepis* (1), *Decaspermum* (1), *Descurainia* (1), *Desmos* (1), *Dianthus* (1), *Dillenia* (2), *Diospyros* (1), *Dipteryx* (1), *Docynia* (1), *Dorema* (1), *Draba* (1), *Dracontomelum* (1), *Dryobalanops* (1), *Durio* (1), *Ehretia* (1), *Elaeagnus* (2), *Elsholtzia* (2), *Eriodictyon* (1), *Eruca* (1), *Ervatamia* (2), *Eryngium* (1), *Erythroxylum* (2), *Eucalyptus* (3), *Evodia* (1), *Farsetia* (1), *Feijoa* (1), *Feronia* (1), *Ferula* (2), *Filipendula* (1), *Fissistigma* (1), *Flacourtia* (1), *Flemingia* (1), *Foeniculum* (1), *Fragaria* (2), *Garcinia* (5), *Gardenia* (2), *Gaultheriua* (2), *Gelsemium* (1),*Geum* (1), *Glycosmis* (1), *Glycyrrhiza* (1), *Gonystylus* (1), *Guettarda* (1), *Gynandropsis* (1), *Heliotropium* (1), *Hemidesmus* (1), *Heracleum* (1), *Hesperethusa* (1), *Hibiscus* (1), *Hippophae* (1), *Holboellia* (1), *Horsfieldia* (1), *Houttuynia* (1), *Humulus* (1), *Hyptis* (2), *Hyssopus* (1), *Ilex* (1), *Illicium* (1), *Inula* (1), *Jasminum* (9), *Juglans* (1), *Jurinea* (1), *Kedrotis* (2), *Laggera* (1), *Lancea* (1), *Lantana* (2), *Laurus* (1), *Lavendula* (5), *Lawsonia* (1), *Lepidium* (1), *Leucas* (4), *Limnophila* (2), *Lindera* (2), *Liquidambar* (1), *Litsea* (1), *Macadamia* (1), *Madhuca* (1), *Magnolia* (1), *Mahonia* (1), *Majorana* (1), *Malpighia* (1), *Malus* (1), *Mammea* (2), *Mangifera* (1), *Matricaria* (1),

Melissa (1), *Mentha* (9), *Meriandra* (1), *Merremia* (1), *Messerschmidia* (1), *Mesua ferrea* (1), *Michelia* (3), *Micromeria* (1), *Microtoena* (1), *Millettia* (1), *Mimusops* (1), *Momordica* (2), *Morina* (2), *Morinda* (1), *Moringa* (1), *Morus* (1), *Muntingia* (1), *Murraya* (2), *Myrica* (1), *Myricaria* (1), *Myristica* (1), *Myroxylon* (2), *Myrtus* (1), *Nardostachys* (1), *Nasturtium* (1), *Nelumbo* (1), *Neolitsea* (1), *Nepeta* (3), *Nerium* (1), *Nicotiana* (2), *Nigella* (3), *Nyctanthes* (1), *Ocimum* (6), *Oenanthe* (1), *Olax* (1), *Origanum* (1), *Osmanthus* (1), *Osmorhiza* (1), *Oxalis* (2), *Paeonia* (1), *Parkia* (1), *Passiflora* (2), *Pavetta* (1), *Pavonia* (1), *Pelargonium* (4), *Perilla* (2), *Perovskia* (1), *Persea* (1), *Petiveria* (1), *Petroselinum* (1), *Peucedanum* (2), *Physalis* (1), *Phytolacca* (1), *Pimenta* (2), *Pimpinella* (2), *Piper* (9), *Pistacia* (3), *Pithecellobium* (1), *Pittosporum* (1), *Pluchea* (1), *Pogostemon* (1), *Polyscias* (1), *Poncirus* (1), *Porana* (1), *Pothomorphe* (1), *Prangos* (1), *Premna* (1), *Prunus* (7), *Psidium* (1), *Pterocarpus* (1), *Punica* (1), *Pycnanthemum* (1), *Pyrus* (1), *Reseda* (1), *Rhamnus* (1), *Rhododendron* (3), *Rhodom-yrtus* (1), *Rhus* (1), *Ribes* (1), *Robinia* (1), *Rosa* (8), *Rosmarinus* (1), *Rubus* (3), *Ruta* (1), *Salvia* (2), *Sambucus* (1), *Sassafras* (1), *Satureja* (2), *Saussurea* (1), *Schinus* (1), *Schisandra* (1), *Schleichera* (1), *Sedum* (1), *Selinum* (3), *Senecio* (1), *Sesamum* (1), *Shorea* (1), *Skimmia* (1), *Sonneratia* (1), *Sorbus* (1), *Spondias* (2), *Stemodia* (1), *Stixis* (1), *Strophanthus* (1), *Syzygium* (3), *Tagetes* (4), *Talauma* (1), *Tamarindus* (1), *Tanacetum* (4), *Telosma* (1), *Thelepaepale* (1), *Theobroma* (1), *Thlaspi* (1), *Thymus* (3), *Toddalia* (1), *Tricalysia* (1), *Trigonella* (2), *Trilisa* (1), *Triphasia* (1), *Valeriana* (3), *Vallaris* (1), *Vanugeria* (1), *Vatica* (1), *Viburnum* (3), *Vigna* (1), *Viola* (1), *Vitis* (1), *Ximenia* (1), *Xylopia* (1), *Zanthoxylum* (6), *Zizyphus* (1).

APPENDIX — II

Families of Flavouring and Fragrant Plants

Name of Family	Total No. of Genera and Species	Total No. of species in Genus
Acanthaceae	(1/1)	*Thelepaepale* (1).
Altingiaceae	(2/2)	*Altingia* (1), *Liquidambar* (1).
Amaranthaceae	(1/1)	*Amaranthus* (1).
Anacardiaceae	(7/10)	*Bouea* (1), *Dracontomelum* (1), *Mangifera* (1), *Pistacia* (3), *Rhus* (1), *Schinus* (1), *Spondias* (2).
Ancistrocladaceae	(1/1)	*Ancistrocladus* (1).
Annonaceae	(6/9)	*Annona* (4), *Artabotrys* (1), *Cananga* (1), *Desmos* (1), *Fissistigma* (1), *Xylopia* (1).
Apiaceae	(21/35)	*Anethum* (2), *Angelica* (2), *Apium* (1), *Bunium* (1), *Bupleurum* (3), *Carum* (5), *Coriandrum* (1), *Cuminum* (1), *Daucus* (2), *Dorema* (1), *Eryngium* (1), *Ferula* (2), *Foeniculum* (1), *Heracleum* (1), *Oenanthe* (1), *Osmorhiza* (1), *Petroselinum* (1), *Peucedanum* (2), *Pimpinella* (2), *Prangos* (1), *Selinum* (3).
Apocynaceae	(6/7)	*Aganosma* (1), *Carissa* (1), *Ervatamia* (2), *Nerium* (1), *Strophanthus* (1), *Vallaris* (1).
Araceae	(5/5)	*Acorus* (1), *Anadendrum* (1), *Homalomena* (1), *Monstera* (1), *Remusatia* (1).
Araliaceae	(1/1)	*Polyscias* (1).
Arecaceae	(6/6)	*Areca* (1), *Arecastrum* (1), *Cocos* (1), *Hyphaena* (1), *Oncosperma* (1), *Raphia* (1).
Asclepiadaceae	(4/4)	*Calotropis* (1), *Decalepis* (1), *Hemidesmus* (1), *Telosma* (1).

Asteraceae	(15/31)	*Achillea* (1), *Ageratum* (1), *Anthemis* (1), *Artemisia* (8), *Blumea* (3), *Chrysanthemum* (2), *Inula* (1), *Jurinea* (1), *Laggera* (1), *Matricaria* (1), *Pluchea* (1), *Senecio* (1), *Tagetes* (4), *Tanacetum* (4), *Trilisa* (1).
Aquifoliaceae	(1/1)	*Ilex* (1).
Barringtoniaceae	(1/1)	*Careya* (1).
Begoniaceae	(1/1)	*Begonia* (1).
Berberidaceae	(1/1)	*Mahonia* (1).
Betulaceae	(1/1)	*Betula* (1).
Bombacaceae	(2/2)	*Adansonia* (1), *Durio* (1).
Boraginaceae	(5/5)	*Borago* (1), *Cordia* (1), *Ehretia* (1), *Heliotropium* (1), *Messerschmidia* (1)
Brassicaceae	(10/12)	*Alliaria* (1), *Armoracia* (1), *Brassica* (2), *Cardamine* (1), *Descurainia* (1), *Draba* (1), *Eruca* (1), *Farsetia* (1), *Lepidium* (1), *Nasturtium* (1), *Thlaspi* (1).
Bromeliaceae	(1/1)	*Ananas* (10).
Burseraceae	(4/7)	*Boswellia* (1), *Bursera* (1), *Canarium* (1), *Commiphora* (4).
Caesalpiniaceae	(3/3)	*Bauhinia* (1), *Cassia* (1), *Tamarindus* (1).
Capparaceae	(4/4)	*Capparis* (1), *Cleome* (1), *Gynandropsis* (1), *Stixis* (1).
Caprifoliaceae	(2/4)	*Sambucus* (1), *Viburnum* (3).
Caryocaraceae	(1/1)	*Caryocar* (1).
Caryophyllaceae	(2/2)	*Caryophyllus* (1), *Dianthus* (1).
Chenopodiaceae	(1/2)	*Chenopodium* (2).
Clusiaceae	(3/8)	*Garcinia* (5), *Mammea* (2), *Mesua* (1).
Combretaceae	(1/1)	*Combretum* (1).
Convolvulaceae	(2/2)	*Merremia* (1), *Porana* (1).
Crassulaceae	(1/1)	*Sedum* (1).
Cucurbitaceae	(3/4)	*Cucumis* (1), *Kedrotis* (1), *Momordica* (2).
Cupressaceae	(3/9)	*Cupressus* (2), *Juniperus* (6), *Tetraclinis* (1).
Cyperaceae	(3/8)	*Cyperus* (6), *Remirea* (1), *Scleria* (1).
Dilleniaceae	(1/2)	*Dillenia* (2).
Dioscoreaceae	(1/1)	*Dioscorea* (1).
Dipsacaceae	(1/2)	*Morina* (2).
Dipterocarpaceae	(3/3)	*Dryobalanops* (1), *Shorea* (1), *Vatica* (1).
Ebenaceae	(1/1)	*Diospyros* (1).
Elaeagnaceae	(2/3)	*Elaeagnus* (2), *Hippophae* (1).
Ericaceae	(2/5)	*Gaultheria* (2), *Rhododendron* (3).
Erythroxylaceae	(1/2)	*Erythroxylum* (2).

Euphorbiaceae	(5/5)	*Antidesma* (1), *Jatropha* (1), *Joannesia* (1), *Phyllanthus* (1), *Sapium* (1).
Flacourtiaceae	(1/1)	*Flacourtia* (1).
Geraniaceae	(1/4)	*Pelargonium* (4).
Grossulariaceae	(1/1)	*Ribes* (1).
Hydrocharitaceae	(1/1)	*Ottelia* (1).
Hydrodictyaceae	(1/1)	*Hydrodictyon* (1).
Hydropyllaceae	(1/1)	*Eriodictyon* (1).
Iridaceae	(3/3)	*Crocus* (1), *Iris* (1), *Tigridia* (1).
Juglandaceae	(1/1)	*Juglans* (1).
Lamiaceae	(27/56)	*Aeolanthus* (1), *Anisomeles* (1), *Calamintha* (1), *Coleus* (1), *Elsholtzia* (2), *Hyptis* (2), *Hyssopus* (1), *Lavendula* (5), *Leucas* (4), *Majorana* (1), *Melissa* (1), *Mentha* (9), *Meriandra* (1), *Micromeria* (1), *Microtoena* (1), *Nepeta* (3), *Ocimum* (6), *Origanum* (1), *Perilla* (2), *Perovskia* (1), *Pogostemon* (1), *Pycnanthemum* (1), *Rosmarinus* (1), *Salvia*(2), *Satureja*(2), *Saussurea*(1), *Thymus* (3)
Lardizabalaceae	(1/1)	*Holboellia* (1).
Lauraceae	(7/17)	*Cinnamomum* (10), *Laurus* (1), *Lindera* (2), *Litsea* (1), *Neolitsea* (1), *Persea* (1), *Sassafras* (1).
Liliaceae	(6/19)	*Allium* (12), *Hemerocallis* (1), *Lilium* (2), *Narcissus* (2), *Polianthus* (1), *Smilax* (1).
Loganiaceae	(1/1)	*Gelsemium* (1).
Lythraceae	(1/1)	*Lawsonia* (1).
Magnoliaceae	(4/6)	*Illicium* (1), *Magnolia* (1), *Michelia* (3), *Talauma* (1).
Malpighiaceae	(1/1)	*Malpighia* (1).
Malvaceae	(3/5)	*Abelmoschus* (3), *Hibiscus* (1), *Pavonia* (1).
Meliaceae	(2/2)	*Azadirachta* (1), *Cedrela* (1).
Mimosaceae	(3/6)	*Acacia* (4), *Parkia* (1), *Pithecellobium* (1).
Moraceae	(5/5)	*Artocarpus* (1), *Cannabia* (1), *Chlorophora* (1), *Humulus* (1), *Morus* (1).
Moringaceae	(1/1)	*Moringa* (1).
Myricaceae	(1/1)	*Myrica* (1).
Myristicaceae	(2/2)	*Horsfieldia* (1), *Myristica* (1).
Myrsinaceae	(1/1)	*Ardisia* (1).
Myrtaceae	(9/14)	*Callistemon* (1), *Decaspermum* (1), *Eucalyptus* (3), *Feijoa* (1), *Myrtus* (1), *Pimenta* (2), *Psidium* (1), *Rhodomyrtus* (1), *Syzygium* (3).
Nymphaeaceae	(1/1)	*Nelumbo* (1).
Olacaceae	(2/2)	*Olax* (1), *Ximenia* (1).
Oleaceae	(3/11)	*Jasminum* (9), *Nyctanthes* (1), *Osmanthus* (1).

Orchidaceae	(2/2)	*Rhynchostylis* (1), *Vanilla* (1).
Oxalidaceae	(2/2)	*Oxalis* (2).
Pandanaceae	(1/2)	*Pandanus* (2).
Papilionaceae	(10/12)	*Butea* (1), *Dipteryx* (1), *Flemingia* (1), *Glycyrrhiza* (1), *Milletia* (1), *Myroxylon* (2), *Pterocarpus* (1), *Robinia* (1), *Trigonella* (2), *Vigna* (1).
Parmeliaceae	(4/11)	*Everniastrum* (2), *Parmelia* (5), *Parmotrema* (3), *Rimelia* (1).
Passifloraceae	(1/2)	*Passiflora* (2).
Pedaliaceae	(1/1)	*Sesamum* (1).
Physciaceae	(2/2)	*Anaptychia* (1), *Heterodermia* (1).
Pinaceae	(6/7)	*Cedrus* (1), *Chamaecyparis* (1), *Cryptomeria* (1), *Picea* (1), *Pinus* (2), *Tsuga* (1).
Piperaceae	(3/11)	*Houttuynia* (1), *Piper* (9), *Pothomorphe* (1).
Pittosporaceae	(1/1)	*Pittosporum* (1).
Poaceae	(6/13)	*Amphilophis* (1), *Anthoxanthum* (1), *Cymbopogon* (8), *Eragrostiella* (1), *Oryza* (1), *Vetiveria* (1).
Polygonaceae	(3/5)	*Polygonum* (2), *Rheum* (1), *Rumex* (2).
Polypodiaceae	(1/2)	*Phymatodes* (2).
Proteaceae	(1/1)	*Macadamia* (1).
Pteridaceae	(1/1)	*Pteris* (1)
Punicaceae	(1/1)	*Punica* (1).
Phytolaccaceae	(2/2)	*Petiveria* (1), *Phytolacca* (1).
Ranunculaceae	(3/5)	*Clematis* (1), *Nigella* (3), *Paeonia* (1).
Resedaceae	(1/1)	*Reseda* (1).
Rhamnaceae	(2/2)	*Rhamnus* (1), *Zizyphus* (1).
Rhodomelaceae	(1/1)	*Laurentia* (1).
Rhodomeniaceae	(1/1)	*Palmaria* (1).
Roccellaceae	(1/2)	*Roccella* (2).
Rosaceae	(11/27)	*Cydonia* (1), *Docynia* (1), *Filipendula* (1), *Fragaria* (2), *Geum* (1), *Malus* (1), *Prunus* (7), *Pyrus* (1), *Rosa* (8), *Rubus* (3), *Sorbus* (1).
Rubiaceae	(8/10)	*Anthocephalus* (1), *Coffea* (2), *Gardenia* (2), *Guettarda* (1), *Morinda* (1), *Pavetta* (1), *Tricalysia* (1), *Vanugeria* (1).
Rutaceae	(15/28)	*Acronychia* (1), *Aegle* (1), *Citrus* (7), *Clausena* (2), *Evodia* (1), *Feronia* (1), *Glycosmis* (1), *Hesperethusa* (1), *Murraya* (2), *Poncirus* (1), *Ruta* (1), *Skimmia* (1), *Toddalia* (1), *Triphasia* (1), *Zanthoxylum* (6).
Salicaceae	(1/2)	*Salix* (2).
Santalaceae	(1/1)	*Santalum* (1).

Sapotaceae	(4/4)	*Achras* (1), *Aisandra* (1), *Madhuca* (1), *Mimusops* (1).
Sapindaceae	(2/2)	*Bl`ighia* (1), *Schleichera* (1).
Schisandraceae	(1/1)	*Schindradra* (1).
Scrophulariaceae	(3/4)	*Lancea* (1), *Limnophila* (2), *Stemodia* (1).
Solanaceae	(4/6)	*Capsicum* (2), *Cestrum* (1), *Nicotiana* (2), *Physalis* (1).
Sonneratiaceae	(2/2)	*Crypteronia* (1), *Sonneratia* (1).
Sterculiaceae	(1/1)	*Theobroma* (1).
Stictaceae	(1/1)	*Lobaria* (1).
Tamaricaceae	(1/1)	*Myricaria* (1).
Theaceae	(1/1)	*Camellia* (1).
Thymelaeaceae	(3/3)	*Aquilaria* (1), *Daphne* (1), *Gonystylus* (1).
Tiliaceae	(1/1)	*Muntingia* (1).
Urticaceae	(2/2)	*Debregeasia* (1), Pilea (1).
Usneaceae	(3/7)	*Evernia* (1), *Ramalina* (4), *Usnea* (2).
Valerianaceae	(2/4)	*Nardostachys* (1), *Valeriana* (3).
Verbenaceae	(4/5)	*Aloysia* (1), *Callicarpa* (1), *Lantana* (2), *Premna* (1).
Violaceae	(1/1)	*Viola* (1).
Vitaceae	(1/1)	*Vitis* (1).
Zingiberaceae	(11/35)	*Aframomum* (1), *Alpinia* (5), *Amomum* (6), *Curcuma* (6), *Elettaria* (1), *Etlingera* (2), *Gastrochilus* (1), *Globba* (1), *Hedychium* (3), *Kaempferia* (3), *Zingiber* (6)

APPENDIX — III

Flavouring and Fragrant* Species under a Genus

Name of Genus	No. of Species	Name of Genus	No. of Species
Allium	12	*Blumea*	3
Cinnamomum	10	*Bupleurum*	3
Jasminum	9	*Eucalyptus*	3
Mentha	9	*Hedychium*	3
Piper	9	*Kaempferia*	3
Artemisia	8	*Michelia*	3
Cymbopogon	8	*Nepeta*	3
Rosa	8	*Nigella*	3
Citrus	7	*Parmotrema*	3
Prunus	7	*Pistacia*	3
Amomum	6	*Rhododendron*	3
Curcuma	6	*Rubus*	3
Cyperus	6	*Selinum*	3
Juniperus	6	*Syzygium*	3
Ocimum	6	*Thymus*	3
Zanthoxylum	6	*Valeriana*	3
Zingiber	6	*Viburnum*	3
Alpinia	5	*Anethum*	2
Carum	5	*Angelica*	2
Garcinia	5	*Brassica*	2
Lavendula	5	*Capsicum*	2
Parmelia	5	*Chenopodium*	2
Acacia	4	*Chrysanthemum*	2
Annona	4	*Clausena*	2
Commiphora	4	*Coffea*	2

Leucas	4	*Cupressus*	2
Pelargonium	4	*Daucus*	2
Ramalina	4	*Dillenia*	2
Tagetes	4	*Elaeagnus*	2
Tanacetum	4	*Elsholtzia*	2
Abelmoschus	3	*Ervatamia*	2
Erythroxylum	2	*Oxalis*	2
Etlingera	2	*Pandanus*	2
Everniastrum	2	*Passiflora*	2
Ferula	2	*Perilla*	2
Fragaria	2	*Peucedanum*	2
Gardenia	2	*Phymatodes*	2
Gaultheria	2	*Pimenta*	2
Hyptis	2	*Pimpinella*	2
Lantana	2	*Pinus*	2
Lilium	2	*Polygonum*	2
Limnophila	2	*Roccella*	2
Lindera	2	*Rumex*	2
Mammea	2	*Salix*	2
Momordica	2	*Salvia*	2
Morina	2	*Satureja*	2
Murraya	2	*Spondias*	2
Myoxylon	2	*Trigonella*	2
Narcissus	2	*Usnea*	2
Nicotiana	2		

***This appendix includes only those genera for which at least two *plant* species are recorded for flavouring and fragrancce.**

APPENDIX — IV

Index to Botanical Names

A

D

E

N

O

P

R

S

Index to English Names

Index to Sanskrit Names

Index to Hindi Names

Other Related Books

ISBN	Title/Author
81-7233-363-3	A Manual of Ethnobotany (2nd Ed.) / *Jain, S.K.*
81-7233-276-9	Betelvine Cultivation & Management of Diseases/ *Chaurasia, J.P.*
81-7233-284-X	Bibliography of Indian Ethnobotany, 2002 / *Jain, S.K.*
81-7233-395-1	Ethnic Aphrodisiac Plants / *Sood, S.K.*
81-7233-412-5	Ethnic Indian Plants in Cure Diabetes / *Sood, S.K.*
81-7233-347-1	Ethnobotany & Medicinal Plants of India & Nepal (Vol. 1-2) / *Singh, V.*
81-7233-182-7	Ethnobotany of Rajasthan (India) / *Singh, V.*
81-7233-137-1	Flora of Andhra Pradesh : Vol. 1-5 / *Pullaiah, T.*
81-7233-179-7	Flora of Upper Liddar Valleys of Kashmir Himalaya (Vol. 1-2) /*Sharma, B.M.*
81-7233-399-4	Floristic Biodiversity of Barda Hills and its Surrounding / *Nagar, P.S.*
81-7233-361-7	Flowers and Vegetables of India / *Sud, R.K.*
81-7233-448-6	Glossary of Medicinal Plants used in Ayurveda / *Saroya, A.P. Singh*
81-7233-450-8	Handbook of Indian Medicinal Plants / *Joshi, M.C.*
81-7233-340-4	Handbook of Legumes of World Economic Importance / *Duke, J.A.*
978-81-7233-552-6	Handbook of Phytopharmacology / *Saroya, A.P. Singh*
81-7233-479-6	Herbal Technology Recent Trends and Progress / *Daniel, M.*
81-7233-360-9	Herbs, Culinary, Medicinal, Aromatic / *Sud, R.K.*
81-7233-477-X	Herbs of Tribal Land, Jharkhand, India / *Jaipuriar, M.K.*
81-7233-373-0	Identification of Common Indian Medicinal Plants / *Naik, V.N.*
81-7233-481-8	Indian Folk Medicines & other plant-based products / *Singh, V.*
81-7233-496-6	Indigenous Knowledge: An Application / *Sahu, T.R.*
81-7233-325-0	Leafy and Edible Plants of North-East India / *Kumar, S.*
81-7233-326-9	Medicinal and Poisonous Plants of India / *Caius, J.F.*
978-81-7233-546-5	Medicinal Plants of India (Vol. 1-3) / *Jadhav, Dinesh*
81-7233-286-6	Monograph on Indian *Leucas* R. Br. (Dronapushpi) Lamiaceae / *Singh, V.*
81-7233-256-4	Monograph on Indian Subtribe Cassiinae (Caesalpiniaceae) / *Singh, V.*
978-81-7233-551-9	Non-conventional Legumes and Useful Plants of North East India /*Kumar, S.*
978-81-7233-533-5	Phytotherapeutic Wisdom of Indian Rurals Aboriginals /*Singh, V.*
81-7233-269-6	Recent Researches in Plant Anatomy and Morphology /*Maheshwari, J.K.*
81-7233-414-1	Sacred and Magico-Religious Plants of India / *Sood, S.K.*
81-7233-344-7	The Poisonous Plants of Bombay / *Kirtikar, K.R.*

For more inquiry visit www.scientificpub.com